THE REVIEWS ARE IN!

"AMUSING . . . WELL-WRITTEN . . .
CAPTIVATING . . . A DISTINCT PLEASURE . . .
COMPARES WITH JAMES HERRIOTT'S WORK. In short, I enjoyed the reading and do not doubt that others would likewise enjoy these stories."

Iowa State University Press

"MEMORABLE . . .
FUNNY, MOVING, SURPRISING . . .
Splendid stories written with wit and imagination from one of America's finest veterinarians . . . Dr. Wasserman is Brooklyn's own James Herriott."

C.C. Kohler, Bookseller,
Dorking, Surrey, England

"My great thanks for the most entertaining stories I've read in a long time . . . Once I picked up the book, I couldn't put it down. It was full of life, humor and compassion and thoroughly engaging. TRULY TOUCHING."

Francesca Anderson, Artist

"Dr. Wasserman created an extraordinary animal practice based on compassion, a true love of animals and an uncanny sense of the bonds people have with them. These stories reflect beautifully, and often humorously, on this man's life and the practice he nurtured. It has been my great fortune to have the opportunity to continue and preserve what he started."

Richard Turoff, D.V.M.

"Dr. Wasserman was my vet for many years and he treats the subjects in his book with the same kindness and generosity he showed to my animals. It is his gift to us all."

Lois Binetsky, Sculptor, Animal Advocate

"His marvelous stories reveal the sensitive and compassionate side of Dr. Wasserman as he spends each day responding to the pet owners and the various ailments of the animal world with great insight into the nature of both the human and animal species. THIS BOOK OF DELIGHTFUL STORIES WILL ENTERTAIN YOU, EDUCATE YOU, AND WARM YOUR HEART."

Harvey R. Fischman, D.V.M., DrPH
Associate Professor, Johns Hopkins University

"I was glad to see Dr. Wasserman put together stories from his practice, dealing with the treatment not only of the usual pets, but on occasion farm animals, horses, and some wild creatures, one of which was mine, a margay, which he has written about in one of his delightful tales. I enjoyed them immensely."

Meg Merrill, Author of Know Your Ocelots and Margays, *Animal Advocate*

"*DON'T* MISS THIS BOOK!
These stories will absolutely captivate animal lovers, especially pet owners. As a small animal vet, they struck knowing and reminiscent chords in me. I loved them."

Christina J. Norton, D.V.M.

"If you love animal stories, 'The Dog Who Met the Queen and Other Stories' is your kind of book. Dr. Wasserman had me eagerly turning to the next chapter all the way through. I found myself at times laughing out loud, weeping, and even holding my breath as the suspense mounted. ABSORBINGLY TOLD BY A SUPERB STORYTELLER . . . RICHLY DETAILED WITH INSIGHT AND HUMOR.

Margaret Jackson, Avon Cat Rescue League, Warks, England

The Dog Who Met the Queen
& Other Stories

Bernard Wasserman, D.V.M.

For a single copy of this book, please contact:
Bennington Press
59 Hicks Street
Brooklyn, NY 11201
Tel: (914) 225-9635 Fax: (718) 522-5445

Book design, production, and cover art by:
The Floating Gallery
331 W. 57th St., #465, New York, NY 10019
Phone & Fax: 212-399-1961
E-mail: FloatinGal@aol.com

Publisher's Cataloging-in-Publication
Wasserman, Bernard, 1920-
The dog who met the queen : & other stories /
by Bernard Wasserman. -- 1st ed.
p. cm.
LCCN: 99-94385
ISBN: 0-9671637-0-6

1. Wasserman, Bernard. 2. Veterinarians--New York (State)--New York--Biography. 3. Pets--Anecdotes. I. Title.

SF613.W37A3 1999 636.089/092
[B] QBI99-897

This book is dedicated to Benjic, my wonderful first cat, to Skippy, my first dog, and to Homer, Homerella, Alexander, Jenny, the incredible Bennington, and to all the great cats and dogs I have lived with and known.

It is dedicated especially to my wife, Bernice, and my sons, Harvey and Andrew, who have shared these delightful animals with me. Also, to Donna Wasserman and my grandson Jake.

And to the memory of my friend and physician, Bert Abel, M.D., a true animal lover.

Contents

Acknowledgements

I want to thank a number of people who read the manuscript and offered their criticisms and comments: my wife Bernice, Lois Binetsky, and my friend and colleague, P.J. Field; Meg Merrill, and my neighbor Fran Anderson, who was one of the first to read the manuscript, and whose effervescent enthusiasm for it was very encouraging, as was Chris Koehler's. And my thanks to Alice Gleason who steered me toward self-publishing.

Most of all, many thanks to my son Harvey Wasserman, whose critical eye and succinct comments were often humorously delivered, and to my editor Amy Yellin, who did a superb job.

Preface

During the 30 years of my veterinary practice, many interesting, strange and humorous incidents took place. I should have taken notes but I never did. When I retired, I decided it would be fun to record these vignettes and stories. Bit by bit, they surfaced from the depths of my memory like ore from a mine. "The Skunk Story," I'd think, and jot that title down, as well as "The Escape Artists" and "The Racehorse Story," etc.

To help get me started, I signed up for an adult education course in writing at Brooklyn College. There were specific assignments, but the student needn't have done them and could write whatever he or she wanted.

I found I was able to recall very clearly events, names of clients, other people and animals involved in the incidents. Most of the names are real. A few have been forgotten or purposely changed.

Ailurophiles

Before I started my own practice, I worked for a veterinarian in Brooklyn. One day we got a telephone call from a man named Jim Cawley. Upset and tearful, Jim explained that the New York City Department of Health had ordered him to get rid of all but two or three of his cats. Neighbors in his apartment building had complained for some time about the horrendous odor in the halls and wanted either Cawley or most of his cats removed. No one knew how many cats he had.

Inspection of the halls and Jim's apartment confirmed that the complaint was eminently valid. Since the cats were wild and impossible to approach, our job was to catch and inject them with a tranquilizer so the ASPCA could put them in carriers and transport them to a shelter.

I took John Conlon, one of the assistants, with me and we headed for Brooklyn Heights, a neighborhood I had never encountered before. We found Cawley's building, and on entering the hallway, the smell of cat feces and urine overpowered us. A short, dark-eyed, unshaven, disheveled man wearing suspenders, worn trousers, a frayed shirt and scuffed shoes opened Cawley's door. A stench many times more powerful than the smell in the hall escaped from the interior of the apartment and emanated from Cawley himself. Once inside the apartment, I knew I would be unable to rid my shoes of the unbearable stench for over a week, and

I would not soon forget the olfactory and visual memory of the scene.

The sight was incredible. The unfurnished apartment held 33 cats scattered in various places and poses. All were apprehensive at the strangers' entry and in a state of hysteria, most scrambled to escape and hide. Cawley had captured the cats from the street, and they were suspicious and fearful of everyone.

The bare oak parquet living room floor held many litter pans piled up to a foot high with cat feces and overflowing with urine. Also on the floor lay a partially gnawed ham turning green with mold, thrown there for the cats to eat. Empty cat food cans were scattered randomly, opened for the cats to eat from and never removed. An accidental brushing against a loose portion of wallpaper stirred hundreds of well-fed cockroaches that were waiting in their hiding place for darkness and a nocturnal banquet. Cawley was not a paragon of hygienic practices.

John and I snared cat after cat as the poor animals frantically tried to climb the walls while I injected them with tranquilizer. We placed them in cat carriers, two in each one. I was heartsick about it, but it had to be done. Cawley screamed, cried, pounded the wall, cursed the Health Department, and cursed God for allowing this to happen. My heart ached for Cawley too. I knew he meant well, thinking he was rescuing and helping homeless cats.

Jim Cawley recovered and eventually, after I opened my own practice in Brooklyn Heights, came in with one cat or another for treatment of various medical problems. He started building his collection again. I advised him about proper management, hygiene and preventive medicine—especially for large populations—but I doubt I got through.

Let me tell you something about Cawley. Despite his slovenly habits, threadbare and malodorous clothing, and lack of formal education, he was an intelligent and good-hearted person. He tinkered with inventions, creating sensible models using makeshift materials and tools. Jim made a miniature model of a rotary motor years before a Japanese auto company developed one. I don't remember his other inspirations, but I recall being impressed. His sole purpose in inventing was to make money to help cats.

Jim was a World War I veteran with a small pension and possibly a modest amount of money from family. He never had a job over the many years I knew him. I suspect he received monetary support from a number of sympathetic women in the animal rescue movement. Women seemed to empathize with his selfless devotion to cats. They accompanied him into the clinic and often paid the fee.

Jim gave a name to every cat he rescued and claimed to remember all the names. Eventually the count got up into scores of cats, approaching 100. He had two chest-type freezers into which he deposited dead cats. I dread to think of the stench that would have resulted had he opened them after the total power blackout that New York City experienced in 1965. I never asked him. I presume he just let them re-freeze.

When the building he lived in was torn down, Jim relocated to the Coney Island section of Brooklyn, where many all-but-abandoned wooden shacks had been used for decades as summer seaside bungalows for New York City residents. He moved his cats two by two in cat carriers, or if he could get some of his friends to help, four or six at a time, traveling by subway. On occasion, he transported a larger number of cats by taxi. He occupied several bungalows and no one bothered him. He soon expanded his animal population to over 150 cats, dogs, at least one rat, and a duck or two. One cannot imagine what the places looked and smelled like.

He came to the office occasionally with a moribund cat,

or for a supply of antibiotics or other medications for the many sick animals he always had. As he aged, he brought the animals in by subway in cat carriers placed into a shopping cart. The trip was at least an hour long each way and required much physical effort. Jim developed a bad limp, and with great empathy, I scarcely charged him for visits or medications.

The mortality rate in his aggregation was very high. A healthy cat he picked up in the street had a much better chance of surviving on its own. Contagious diseases such as severe respiratory infections, feline leukemia, infectious peritonitis, and others were rampant due to open exposure, urinary tract blockages, and abscesses from bite wounds. Cats were neither altered nor segregated, so new litters came along constantly, with the kittens having zero chance of survival.

I tried to convince him not to harbor any more cats and to reduce the numbers by attrition or possibly to adopt some out, but to no avail. Jim slept on a mattress on the floor amidst this maelstrom of animal life and death. On one occasion, he told me he hadn't slept the previous night because a rat had taken over the mattress and wouldn't let him on it!

During my final years of practice, I did not see Jim. I wondered about him but couldn't make contact. He never had a telephone or an address. I suspect he died.

Not all extreme ailurophiles—and there are quite a number with large cat collections in the city—are derelict in proper care and management, as was Jim. I have a friend with a good-sized colony, probably 100 cats, but they are cared for as properly as are individual cats in a household. They are all immunized against the routine diseases and are promptly treated at home or brought to a veterinarian when a medical or surgical problem arises. Litter pans are routinely and frequently emptied, and the cats are separated by gender. Fresh water is available at all times and the cats are fed a nutritious diet. They are all sleek, happy and good-natured. On rare occasion when a new one must be added, it is quarantined for a couple of weeks to make certain it isn't harboring a contagious disease.

Poor Jim Cawley devoted his whole life to cats and felt he was doing something good and noble. I liked Jim.

Not long after I opened my own practice in Brooklyn Heights, the old red brick building where Cawley had lived became famous for something other than Cawley's cats. A top Russian spy had been operating out of an apartment in it for years and he had just been captured. Telescopes, radio equipment—the works. His name was Colonel Abel and he was later exchanged for pilot Gary Powers, whom the Russians had shot down and captured when he was flying a U-2 spy plane on a mission over the Soviet Union. Remember that?

A Practice Builder

Rather than work for a long period of time for another veterinarian or buy an existing practice from one who was retiring, I decided to start my own. I found a lovely area in my hometown that, since the demise of an old practitioner, had existed without a veterinarian for several years. An empty storefront in a unique, old federal-style building caught my eye. Thinking to rent the store, I contacted the owner and was informed that the entire building was for sale. Luckily, the zoning was right for my purpose. After the loans, mortgage, and renovations, I was ready to open.

In those days, professional ethics rules forbade advertising except for formal notices in three consecutive issues of a local newspaper. No circulars, mailings or extensive advertisements. Even the listing in the telephone directory was proscribed: fine print and bare essentials. So because of the rules, once I opened my door, ready and eager to get started, I played the waiting game. Every so often (though not frequently enough to suit me), someone brought in a pet. I was somewhat discouraged, and wondered how long it would take until I could earn a living.

Then, in walked Mildred Megargee.

Mildred came through the door of my veterinary clinic just three weeks after I'd opened. She was a tiny, alert, gray-haired, blue-eyed woman, and she had in tow a small dog that was not as eager to see me as was her owner.

Heidi, her eight-year-old Dachshund, had a history of difficult urination with an increasing amount of blood in her urine. On examination, it was quickly apparent that two large calculi (kidney stones) had passed into her urinary bladder.

"Why hasn't this been taken care of before now?" I asked Miss Megargee. "These are so obvious and your dog has been in great discomfort. This has been going on too long." I was very annoyed.

Tears welled in her eyes and she looked embarrassed. "I really couldn't afford to see a veterinarian or to pay for whatever treatment would follow."

Now I felt embarrassed—and excited. I had yet to do the first surgical procedure in my own practice.

"Your dog needs an operation called a cystotomy to remove the stones from her bladder."

"I'm afraid I'll have to put Heidi to sleep. There is no way I can afford the surgery." Her tears fell.

I knew that I would do this operation without payment if necessary. I was so eager to break the ice, surgically speaking. "We'll work something out," I said. "Perhaps small payments over a long period of time?"

"I'm afraid I can't even handle that, Doctor." Copious tears rolled down her face, catching her steel-rimmed eyeglasses. Instead, she made her offer.

"Do you have a pet whose portrait you would like to have painted?"

Miss Megargee was a rare type of artist—an animal portraitist. She portrayed in oils dog show champions and pets of prominent people including the Pugs of the Duke and Duchess of Windsor. Some of her paintings previously appeared on the cover of the magazine Dog World and other publications. Commissions were rare and she had fallen on hard times.

Should I embark on a barter system in my practice? I

really needed the money to pay loans for the building renovation, the mortgage, and the expenses indigenous to operating the practice, not to mention eating. Would this set a bad precedent? "Oh, the new vet will swap his skills for your anything."

But the more I considered it, the more appealing it became. Why not, I thought? I could determine in the future, in each instance, whether I wanted to barter or not. I was anxious to do the surgery, and besides, I would love to have a painting of my wonderful Siamese cat Benjie. I agreed to the barter: her artistry for the surgery.

The operation was prompt and the patient made, as the saying goes, an uneventful recovery. Each calculus was the size of a chestnut, the flat sides facing each other. When Miss Megargee brought Heidi back in to have her sutures removed, she asked if she could keep the two chestnut-sized calculi for a while. Of course.

She came to my apartment above the practice shortly thereafter to look at Benjie's eye color and physical characteristics. I gave her a color slide of Benjie as well to help her in her painting. When Mildred invited my wife and me to her home to see some of her work, we observed her excellent portraits, which made me all the more pleased with my arrangement. Before we left, she loaned me a beautiful framed oil painting of a Cocker Spaniel to hang in my waiting room—she had painted it for the cover of a magazine. Years later, Mildred told me to keep it.

In the ensuing weeks as the practice began to grow, periodically a client would mention the stones I had removed from Mildred Megargee's dog. These clients grew rapidly to a considerable number. It occurred to me that Miss Megargee, a woman who loved chatting, was, without solicitation, doing a priceless public relations job for me! While walking Heidi several times daily, she engaged fellow dog walkers in conversation and showed them the stones that she carried with her at all times. I couldn't have hired a high-priced advertising agency to do a better job. I was grateful.

In time, she brought back the stones and Mildred delivered a lovely oil portrait of my cat—a near-perfect likeness that also captured his unique personality.

Heidi lived another eight years, to age 16, and I saw her for other ailments and vaccination boosters many times during that period. I then cared for her successor, a male Dachshund, until Mildred moved to another city. I never charged Mildred a fee. Her initial visit resulted in my first surgical procedure in the practice, a peerless advertising campaign that, I am certain, accelerated the rate of growth of my practice, and gave me a wonderful remembrance of my dear cat Benjie.

I told Mildred I wanted to keep the stones as a souvenir of my first surgery in the practice. When I retired and was gathering the numerous mementos collected during 28 years of practice, I ran across a jar holding the Heidi stones. I held the two white objects in my palm, rubbing them together and recalling that first momentous visit of Mildred Megargee and her pet.

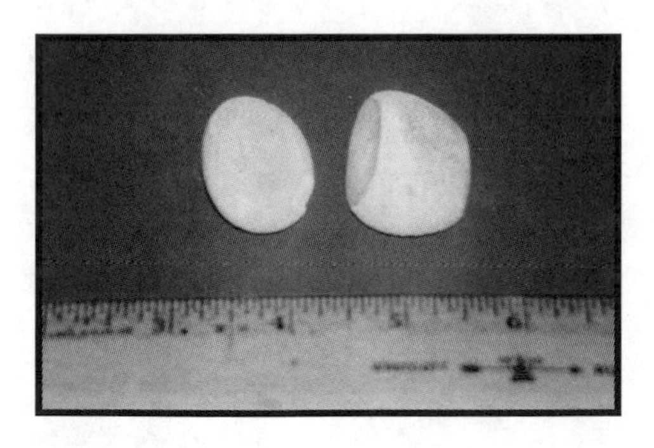

Cystic Calculi from Heidi's bladder.

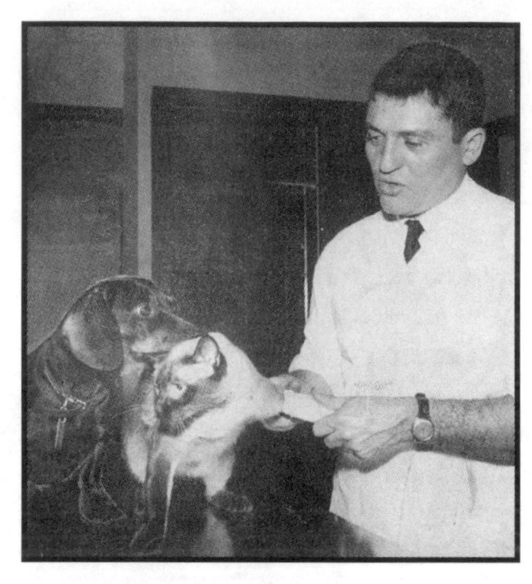

Above: May 1957: Applying a bandage to Benjie while a Dachshund, recovering from a slipped disk, complicates the procedure. Below: The Heights Hospital.

Capote

My sister Gert handed me a slip of paper with the name and telephone number of a client who'd come into my new veterinary office while I was out getting some supplies. I glanced at the name on the slip. Oh my God! Truman CAPOTE came in here? I had read the beautiful and sensitive writing of this young author and kept his work in my collection: *The Grass Harp*, *Other Voices, Other Rooms*, and *A Tree of Night*. To meet him exceeded my expectations.

Some veterinarians, because of the location of their practices, have famous clients. I know colleagues in Manhattan who treat the pets of actors and singers in Broadway shows and musicians known throughout the world. My practice was located in the Brooklyn Heights section of Brooklyn, also a part of New York City, where a disproportionate number of writers have elected to live throughout the years due to its quiet atmosphere and its proximity to the East River and Manhattan. Unlike Manhattan, whose tall buildings make one feel as though one lived at the bottom of a canyon, "The Heights" (as it is called by most residents) has a predominance of 19th-century brownstones and carriage houses, and a small number of six- to eight-story apartment houses. It has the open feeling of a smaller town and a history dating back to the Revolutionary War. One could easily walk over the bridge to the Wall Street financial district or the Chinatown section of Manhattan. The Heights was referred to as New York's first suburb.

The office was still in the process of being set up, and my family was graciously helping out, with Gert doing a bit of receptionist work. My father Chaim, a wonderful painter and decorator, was putting some finishing touches on the waiting room, and so was present at the time when eagerly I telephoned Mr. Capote and made an appointment to see him. He had two dogs: a Kerry blue terrier and an English Bulldog, both males. The latter had been given to him by the cast of the movie *Beat the Devil*, for which he wrote the screenplay. At this time, I do not remember which dog he brought in or the complaint. I do, however, recall the strange-looking man. He was blond, short and a bit pudgy. He wore wire-frame eyeglasses over deep, penetrating blue eyes that sent messages to a mind capable of sizing me up completely in quick time. Most of all, I remember his voice: supersoft, boyish, high-pitched. Strange at first, until one became accustomed and connected the voice to the man. Then it sounded right.

He lived with his friend Jack Dunphy on Willow Street, three blocks from my office. Over the next six months or so, he came in a number of times with his animals.

One night about 10 p.m. my doorbell rang (I lived above the office). It was Jack, very distressed, asking if I could come over to their apartment immediately.

"Kelly, our Kerry blue terrier, collapsed in the street while I was walking him. I carried him home and he is lying on the floor in agony."

I went without delay.

A prostrate dog greeted me in the entrance hall of the apartment, howling in extreme pain. A concerned and worried Truman, near to tears, lay on the floor alongside the dog, stroking and speaking softly to him. Jack hovered nervously over the two.

An examination of the elderly dog quickly revealed the problem: a double perineal hernia, in which Kelly's urinary bladder protruded into the right hernial bulge. In older male dogs, sets of muscles on both sides of the rectum sometimes lose their strength and elasticity and spread apart. The muscles form the posterior or pelvic boundary of the abdominal cavity. In aging males these muscles may weaken considerably, and organs like the small intestine and the urinary bladder may force their way between them, resulting in a large lump under the skin on either side or both sides of the anus. In Kelly's case, the urinary bladder had a fair amount of urine in it and created a large lump along the right side of his rectum. The left side had a hernia as well, which seemed to contain only fatty tissue.

With my hand, I enveloped the larger hernia containing the bladder and gently and gradually manipulated its contents back into the abdominal cavity. Kelly, afforded instant relief from his agony, stood up, wagged his stump of a tail, and immediately expelled at least a quart of urine onto an expensive Oriental rug. What would have been a disaster to Truman under any other circumstances was a joy worthy of celebration. Kelly now asked for food and water and was taken care of, and Truman, taking care of us, brought out a bottle of Courvoisier cognac and three snifters to celebrate Kelly's relief and to soothe frazzled nerves.

This treatment was only temporary; Kelly needed surgery to close the weak areas permanently. We made an appointment for the following day.

As the grateful owners offered me a drink, I had a chance to observe the apartment. Truman was a prodigious collector of all sorts of small items, fine and in good taste: ceramic animals, paperweights, shells, marble and ceramic eggs, small decorative boxes, etc. Virtually every square inch of end tables, cocktail tables and chests was covered. I saw the Victorian couch on which he was photographed in his twenties for the dust cover of one of his earlier books. There were plenty of filled bookcases.

In the entrance hallway, propped against a wall, was an original Toulouse Lautrec poster, unframed and with white drippings on it from a ceiling that had just been painted. I asked him about it. The curator of the Museum of Modern Art bought it for him and he had no place to hang it. Not only was the poster very large, but almost all his wall space was covered with paintings, prints and other wall hangings. The poster shows a cabaret dancer in a yellow skirt kicking her leg high in a can-can dance. In the right foreground is the face and arm of a musician, his large hand gripping the neck of a bass viol. "Jane Avril" is printed near the top, and in the lower margin is the address of his studio and the signature T. Lautrec '93. It was an advertisement for the dancer appearing at the Moulin Rouge. I'd seen pictures of this impressionist poster in various art books and had very much liked it. Although he ran off many hundreds of them in 1893, few have survived and they have become quite valuable.

An anxious Truman dropped Kelly off for surgery the following day and I assured him that most likely all would go well. The anesthesia, the repair of both hernias, and the castration (which is called for in perineal hernia cases, since it is felt that altered male hormones in aging dogs are responsible for the weakness) took about two hours. All went well and I was ably assisted by Ivy Reade, who graciously volunteered her services.

On the following day, a happy Truman came to visit the hospitalized Kelly. He carried an inscribed and autographed copy of on of his books, *The Muses are Heard*. It was an account of his trip to Russia, accompanying the "Porgy and Bess" opera troupe in their tour of the Soviet Union at the request of the U.S. State Department. His inscription to me was personal and gratifying. He visited for awhile with his pet, who was, of course, happy to see him.

One month later, returning from lunch, I saw the author walking both dogs on leashes. I brought up something I had been meaning to ask for some time, but felt hesitant because of his celebrity status.

"Would you and Jack like to come up for dinner some evening? My wife is a marvelous cook."

"We'd love to," was his immediate response. I called him later and set a date.

Bernice prepared a delicious though basic "meat 'n taters" dinner of pot roast, potato pancakes and applesauce, salad, and for dessert, lemon chiffon pie. All the recipients of this repast thoroughly did the meal justice. The evening proved to be most delightful and congenial. Truman and Jack treated us to stories of meetings and conversations with world-renowned people in literature, the arts and politics. One story Truman told had him sitting in the White House talking to his good friend Jacqueline Kennedy. They had talked for several hours following dinner when in walked the President, obviously sleepy and wearing an old bathrobe and worn house slippers. "Truman," he said, "get your ass out of here. Do you know it's 3 a.m.?" Another involved his lunch at Buckingham Palace with Queen Elizabeth and the Queen Mother. I read somewhere that Truman sometimes told of events that never actually happened, but Bernice and I like to think these did.

Moderately tall, blue-eyed, red-haired, and with a florid and extremely freckled complexion, Jack was a natural, iconoclastic comedian. When he told stories, no creed, race, or ethnic group was spared his humorous barbs, including his and mine. He had us roaring with laughter over his version of incidents, which had occurred on one of their stays in Italy.

When the evening came to a close in the wee hours, we felt too charged up to fall asleep.

Some months later, a client called my attention to a story in an issue of Holiday magazine (a now-defunct publication). It was called *A House in the Heights*, written by Truman Capote, and was largely about why he loves living in Brooklyn Heights. Among other things, he related how people from other neighborhoods abandon animals in the Heights, knowing the high percentage of animal lovers there. In the story are a few lines about me:

> "Astonishing really, the amount of lost strays who roam their way into the neighborhood, as though instinct informed them they'd find someone here who wouldn't abide being followed, but would instead, lead them home, boil milk and call Dr. Wasserman, 'Bernie,' our smart-as-they-come young vet whose immaculate hospital resounds with the music of Bach concertos and the barking of mending dogs."

Imagine how thrilled I was to see my name mentioned in a national magazine, and especially by so famous a writer. I even heard from a client who saw the article in Paris. The story was subsequently selected for three hardcover books of short stories: *Selected Writings of Truman Capote*, *The Dogs Bark* and another collection of Capote stories. And recently it appeared in a new collection, *The Brooklyn Reader*, which has stories about Brooklyn by current or former Brooklynites.

A couple of months went by during which time Truman and Jack visited the clinic several times for routine vaccinations for the dogs, and to bring in their new female calico kitten, Diotema. Jack rescued Diotema from the sea in

Greece after a boy tossed her from a rocky promontory. He dove off a cliff and saved the two-month-old kitten from certain death.

Jack was a good athlete. He skied all winter and was a dancer earlier in his career. He wrote several interesting items, including a play that opened and soon closed on Broadway, a couple of novels, and a book called *Dear Genius* about his life with Truman in which Jack has two personae—himself as Jack's friend and himself as a priest.

On one visit to my office, Truman invited us to dinner, which he offered to prepare himself. We were happy to accept. After drinks we were served oefs en gelée, boeuf Bourgignon and fruit and cheese for dessert—all beautifully prepared, delicious as well as pleasurable to the eye. He later confessed to Bernice that he was very nervous making dinner for her, knowing she is a marvelous cook. Once again, we spent a wonderful evening full of laughs and conversation.

Months passed and once again, Truman and Jack returned to our house for dinner. I invited a close friend of mine, Harvey Fischman to meet them.

Dinner being over, we sat around and talked, and once again, Truman and Jack charmed and intrigued us with a myriad of interesting stories about events in their lives. Somehow, Truman brought up the subject of the Toulouse Lautrec poster, which still leaned against the wall of his hallway.

"I'll tell you what. If you'll hang the poster on your waiting room wall where lots of people will see it, I'll let you have it on an indefinite loan."

"I'd love to have it," I said, "but it's so large, it would almost reach from floor to ceiling. It would also have a bench in front of it so a good portion of the poster would be hidden." I also thought it might be damaged or possibly stolen, so I thanked him and reluctantly declined his wonderful offer.

Hours and several drinks after dinner, Truman brought up his first meeting with Bernice. "I didn't know you were married," he said. "Jack and I discussed it and decided she had been away at the University of Wisconsin and had come in for a couple of weeks in February for winter vacation." His fertile imagination even conjured up a specific university!

"I wasn't married until a couple of days before you met her," I said. "We were unable to get out of New York City for a honeymoon because of the snowstorm." There had been a record snowfall the day of the wedding and a good many invited guests couldn't attend.

"You should have invited us to the wedding. We're good gift givers," said Truman.

"Well, it's never too late to give a gift," I said, tongue-in-cheek.

A moment of pondering and then: "Tell you what. If you'll come over to the apartment, we'll give you the Lautrec poster as a belated wedding gift."

Stunned and delighted, I made feeble protests, "It's too much, you must want it, are you sure?" But he insisted. He genuinely wanted us to have the poster.

We all walked back to their apartment just a few blocks away on Willow Street—Jack, Truman, Bernice, my friend Harvey and I. We settled down there with more drinks and conversation, the topics long forgotten, and when the time approached 2 a.m. Bernice, Harvey and I took leave with "Jane Avril." Balancing this very large, unwieldy poster mounted on heavy cardboard, we made our way through the deserted Brooklyn Heights streets. After so heady an evening (punctuated with how many cocktails?), we must have appeared a strange giggly group to the few passersby we encountered.

The poster was cleaned, framed and hung. Periodically, we received books directly from the publisher as they came out, sometimes signed by Truman, sometimes with an enclosed card, which read: Compliments of the Author: *Breakfast at Tiffany's*, *In Cold Blood.* One time we were given a large, beautiful book of celebrity photographs by Richard Avedon, with Truman Capote's comments about the subjects. We also received a record, *A Christmas Memory from Breakfast at Tiffany's*, read by the author.

Truman and Jack eventually moved out of the neighborhood to a condominium in the United Nations Plaza in Manhattan. They also spent much time at their houses in Bridgehampton, one for Truman, one for Jack. And since Truman was frequently on the West Coast, we scarcely saw or heard from him for a long period.

One day in my office, I received a frantic call from Truman. His new English Bulldog (Bunky had died a few months earlier) had jumped out of Truman's convertible automobile while he was driving on the Long Island Expressway the previous week. The dog was struck by another car, causing a deep 10-inch-long gash on the right side of his chest. It was sewn by a nearby veterinarian, but now the entire wound suddenly opened up and the vet was unavailable. I told him to come over immediately, and since my clinic hours ended as he arrived, I was able to start resuturing the wound while he visited upstairs with Bernice, and met my one-year-old son Harvey, whom he had not yet seen.

When the dog awoke from the anesthesia and dog and owner were ready to depart, he asked what he owed me.

"Nothing," I said.

"You should charge me even more than others. I'm rich."

After I said something to the effect that he should learn to accept as well as give gifts, he relented, thanked me, and left.

We never saw him again, but followed his unfortunate decline, reading about him in the newspapers periodically and watching sad, pitiful television interviews.

He died on August 23,1984.

Houdini and Other Escape Artists

One of the biggest nightmares a practicing veterinarian can experience is the escape of an animal in his care. Until the pet is tracked down and returned, the veterinarian goes through torment. He must find out how and why it happened and take preventive measures. Despite the utmost in precaution and carefulness, this sometimes occurs. Fortunately, the following incidents are actually all of such occurrences in my 30 years of practice.

The first one happened just after I had opened my office. I had spayed a cat named Mitzi who was seven months of age, grey and white, with yellow eyes, and a very petite wiry build. She was in an upper cage of a new set of kennels, which had vertical bars about one and one half inches apart on its doors. The window next to the kennel was open two inches from the bottom and awaiting a screen, which was to be delivered later that week.

The morning after the surgery, I went into the kennel room to check on my patient. She was gone! It felt like a blow to my heart. Since the kennel door was still closed, I figured that she must have squeezed through the bars of the door, gotten her paws onto the windowsill and slithered through the two-inch window opening. Nothing amazes me about the feats such fluid, supple animals can accomplish.

The spaces cats can maneuver into, through, and under would astonish anyone.

Mitzi's escape occurred on a Friday preceding the three-day Fourth of July holiday weekend, and because Mitzi's owners had left town, she was due to stay until Tuesday. Ordinarily, she would only have stayed overnight, so at least I had several more days to find her.

I went upstairs and told my wife about this dreadful development. Hoping Mitzi chose not to go out the window and was still in the hospital, we made an exhaustive search of the premises, but she was nowhere to be found. I truly felt that our chance of getting her back was zero.

Together, Bernice and I took to the streets. We went up one and down the next, crisscrossing every part of the neighborhood, searching every alley, yard, garden and cellar entrance calling out "Mitzeeee, Mitzeeee!" We looked through every bush and cellar window. One kind woman reached into her purse and proffered a dollar bill as we were looking into some open garbage cans. Passersby glanced at us quizzically and then grimaced at each other. Some stopped to talk. "Did you lose your child?" "Who is this Mitzi?" At midnight, empty-handed, we quit.

I couldn't sleep. The poor, bewildered kitten that had been sheltered and loved was now lost and hiding in strange places. How do you explain this to an owner? Sure, it was an extremely freakish set of circumstances, but the end result was the same—no cat.

We spent Saturday and Sunday scouring the neighborhood in the same manner, asking people if they'd seen a cat of that description. At night we searched with flashlights, but all we did was awaken several sleeping vagrants who asked for a handout.

Monday. Another safari through the streets of Brooklyn Heights. No luck. Hope waned by the hour and, full of dread, I was resigned to telling the owner what happened.

Early that evening, my doorbell rang. At the door was my neighbor and client from Willow Street, one block away, Mrs. Von Bechstein.

"Dr. Wasserman, there is a small cat on my window sill. We fed her and while handling her, we noticed she had stitches on her belly. We thought you might know to whom she belongs."

I was speechless, elated and a bit unbelieving. Could it be? Quickly, I followed Mrs. Von Bechstein to the backyard of her ground-floor apartment. Mitzi was there on the window ledge, petted by Mr. Von Bechstein. Hugging Mitzi, I felt as though black clouds had suddenly been blown away and the sun shone brightly. She was in good physical shape, totally unaware of the torture she had put me through over the past three days.

The following day her owners came in to collect her.

"How's she doing, Doc?"

I smiled innocently. "Fine," I said.

* * * * *

In the early days of my practice, we groomed dogs. At the time, we had a bathtub in the kennel room, which is at the very rear of the clinic. I would schedule a few dogs for clipping and bathing, and one day a week, a professional groomer would come in and take care of them. For an animal to be able to escape, three doors would have to be open at the same time—kennel room door, door from examining room to waiting room, and the door from the waiting room to the street—a distance of 40 feet. The chance of all three doors being open at the same time while a dog was free in the hospital was incredibly unlikely.

Pat Bellaccio, the groomer, had on the table a six-month-old male Afghan hound, which he was combing and brushing, when he was distracted for a moment by someone at the kennel room door. He took his hands off the pup for a moment to see who it was, and Omar bounded through the door as he opened it. The receptionist opened the examining room door to call a client in from the waiting room at the same time a new client was entering the waiting room from the street. Omar, a flash of greased lightning, was off and away! I watched this disaster in disbelief, immediately made a fast apology to the client whose cat I was examining and dashed out into the street. I saw the tan and black rear end of a tall dog disappearing around the next corner so I gave chase.

Anyone who knows about lean, long-legged Afghan hounds is aware of how fast they can run. Luckily, Omar ran a bit of an erratic course. The chase scene would have been chock-full of laughs if it had been in a movie. He ran into a yard and out, or up steps and then down again and then he took a straight path. He rounded several more street corners; I was fortunate enough to keep him in sight. Omar and I could not keep up the chase much longer. He stopped on a small patch of grass in front of an apartment building. Slowly, I walked towards him, softly calling his name. He stayed and watched me until I was about to reach out for him—and then he bolted. Too late! I leaped on him. He went down on the grass and I hugged him to me. Holding him with my left arm, I slid my belt off my trousers and looped it around his neck. We walked the 6 blocks back to my office at a leisurely pace. We were both played out.

Shortly after this incident, we had the bathtub removed from the kennel room and gave up grooming. This event was not the sole reason, but it helped.

* * * * *

Ready for another one?

My friend Joe was going away for three weeks and asked if I could keep Muffin, his four-year-old grey and black female tabby cat for him. I was glad to do my long-time pal a favor, and Muffin was placed in a large kennel meant for a big dog. She got her pan of kitty litter, water, toys and canned feline delicacies Joe had brought for her, which rivaled in cost (and taste) some human gourmet foods. Joe and family were dotty over this cat.

The staff of the clinic knew this sweet-natured cat since kittenhood. We exchanged pleasantries with her when passing her kennel and dispensed hugs when time permitted. Muffin seemed very contented. All that the accommodations lacked to make it the Plaza Hotel for cats was a view of Central Park. The kennel room was even air-conditioned. One would think a pussycat in such congenial surroundings would want to stay.

Well, Muffin didn't. One morning we found the kennel door open and Muffin gone. At first we didn't worry. She could not have gotten out of the kennel room, or the building, since all the doors were locked. We were sure to find her in some corner, or behind some stacked cases of animal food or piles of newspapers we used for lining the kennels and kept in an open closet off the kennel room. One by one, all the possibilities were checked and eliminated.

Since we had taken great pains to remove all unconventional modes of egress and places of concealment following the previous anxiety-producing incidents, we were now stumped. My heart once again dropped into my abdominal cavity. In an effort to solve Muffin's disappearance, I sat down with my staff to see if something had been overlooked. It hadn't. All doors had been securely closed. Could the kennel door have been latched improperly, so it opened when the cat leaned against it? Yes, that was a probability.

Once again our search covered the entire hospital. We left no spot unexplored, but were still without a happy result. Joe and his family would return in two weeks, a nice long time for me to worry about the well being of his sweet cat. I visualized a number of calamitous, sorrowful scenarios as I informed the family that their beloved Muffin had mysteriously vanished.

About a week after the disappearance, while examining a patient, I heard faint, intermittent scratching noises coming from the ceiling. When the examination was finished, I asked my assistant Ronnie to listen, and (of course) there were no noises for some time. But then I heard it again! There was no mistaking it.

The entire clinic had a dropped ceiling with a space of about one-and a-half feet between the paneling and the regular ten-foot-high pressed tin ceiling. When I acquired the 142-year-old building, all the exposed plumbing, heating and gas pipes hung from the ceiling or were attached to the walls close to the ceiling. We covered the unsightly pipes by building a lower ceiling. As it occurred to us that the scratching noises could be Muffin, we examined the ceiling, room by room to see if there was some point of entry. When we checked the storage area off the kennel room, we discovered a foot-square ceiling panel that had never been installed. Once out of her kennel, Muffin must have scampered up cases of pet food and shelves, and made a short leap through the opening. Since I was smaller and lighter than Ronnie, I climbed a ladder, removed an additional panel for more room and stuck head and shoulders into the pitch-black opening.

Shining the beam of a flashlight in an arc, I saw a pair of luminescent eyes reflecting the glow. Muffin crouched ten feet away from me. I called to her softly and repeatedly cajoled her into approaching, but she seemed to be glued to the spot—she didn't move. Ronnie handed over a noose-like loop on a long handle that we use to control viciously behaving dogs and cats without hurting them (and to prevent them from hurting us). However, even with the noose and my arm's length combined, I could only reach out six feet. I had to crawl into the foot-and-a-half space between ceilings onto flimsy insulating tiles held by a thin metal framework and hangers. Would it hold me for the four or five feet I needed to get within reach of her? Would she run off just as I got in catching distance? If she ran, I couldn't possibly keep crawling after her in that dark, unbearably hot and fragile environment. I was prepared to dismantle the whole ceiling if necessary.

But luck was with me, and she stayed rooted to the spot as I crawled towards her. I got the loop over her head, gently tightened it, drew her back with me to the opening, and handed her to Ronnie.

How did she survive a week up there without food or water? Muffin was quite overweight at the start of this misadventure. In the absence of food and water, her body had broken down some of her fat deposits, and one of the breakdown products of fat is water. She was in reasonably good shape, though very dehydrated. We remedied this with intravenous and subcutaneous fluids and her favorite foods. A longer period of deprivation could have caused some serious permanent changes including neurological damage and death.

Three weeks after he left Muffin, a smiling and well-tanned Joe came to collect her, accompanied by one of his two daughters, Melissa, age six.

"Well, Bern, how's my third daughter? Did she miss us as much as we did her?"

"She's fine, Joe, and I know she missed you all. By the way, we took some weight off her and I want you to try to keep it off. Don't overfeed her. Obesity is no good for man or beast."

I decided to relate the saga of his disappearing pet at some future date.

* * * * *

When we were still boarding cats, the Lindemanns brought Ambrose in to stay for several days.

"Watch out for him, Doc. He can escape from any kennel," cautioned Adrienne Lindemann. "He's done it several times in other boarding places." In fact, she told me when she lived in Manhattan she boarded her cat at the clinic of a veterinarian who also lived above the office. Dr. Asedo had awakened one morning to find Ambrose in bed with him.

"Not to worry. We have very good latches on our kennels and we'll keep an eye on him." A nutty owner, I thought.

Ambrose, a grey and white male domestic short hair, about four years old, was placed in an upper-tier kennel. The attendant made sure the door was securely latched. The latch consisted of a metal horizontal bar on the door, the end of which fitted into a metal sleeve on the frame. Insert into sleeve to lock, lift up to open. Not complicated, but it required a bit of maneuvering to operate. We entered the kennel room the following morning to find emptiness in the kennel where Ambrose resided the previous day. He really was an escape artist! He had figured out how to lift the bar high enough to disengage it from the sleeve and open the door. We found him easily, resting atop some cases of food, calmly observing our astonishment, as if to convey, "My owner told you about me. You shouldn't be surprised." We returned Ambrose to his quarters, and locked the kennel door as usual. He was checked whenever someone had occasion to enter the room. At night, before closing the clinic, we not only latched his door but also bound the latch together with twine.

Let's see this kitty work his way out of that, we thought.

Morning . . . start of a new day. We were greeted by an open, empty kennel, twine hanging down from the latch—and you-know-who, lying atop a case, once again watched our display of incredulity, a smug look on his face.

After that last feat, we decided not to take more drastic steps to keep him enclosed. If he liked to play this game and was so successful at it, he deserved his fun. We also realized the Lindemanns were not nutty animal owners, and Ambrose was thereafter known as Houdini.

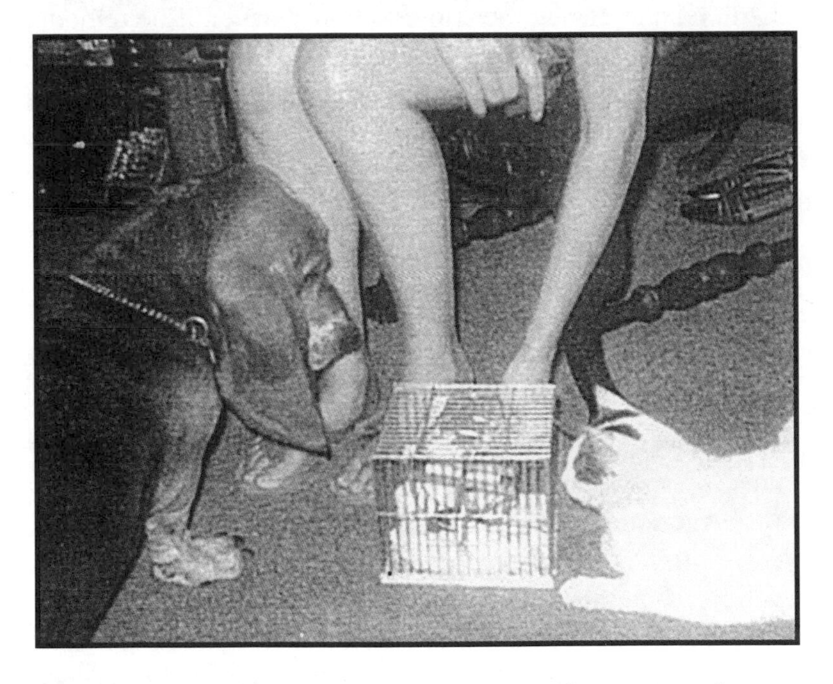

Ambrose with friends, including mouse he captured.

Blackout

Recently, I learned that Bill Arnholt was in the hospital: he was not expected to live. I was very saddened by this news, and decided to telephone him. He was pleased to hear from me, and during the conversation, I asked if he remembered the disaster involving his cat Samantha on the night of the blackout. He did. I told him I was writing a story about it and he liked that. Sadly, Bill died a few weeks later.

* * * * *

Sixteen minutes, eleven seconds past 5 p.m. November 9, 1965. To people who do not live in New York City (and to many who do and are too young, or have forgotten), this date has no significance. But I remember it. It was the night of the great power blackout in New York City. From whatever cause, there was no electricity available in the city. No lights, no refrigeration, no anything having to do with electricity.

At first I was tempted to cancel office hours that evening, but then, we rounded up whatever candles we could get and placed them in strategic positions in the waiting room and the examining room. It was very inadequate and as I was about to close, there was a knock on the door. It was Bill Arnholt, a breeder of Siamese cats who lived in the neighborhood.

"Doc, I've got a problem. Samantha here had kittens a

few days ago and now she seems very sick. She has something hanging out of her rear end."

He put the towel-wrapped bundle on the examining table and uncovered his cat. She was very quiet but in obvious distress. One look confirmed that she had a prolapsed uterus; it had turned inside out and was protruding a good four inches outside her body! Further examination showed her body temperature to be several degrees below normal—a poor sign. She had been in this state all day and Bill only discovered her when he had come home from work. The uterus was extremely swollen—the tissues engorged with blood—and was not in a state to be easily manipulated back into the body cavity.

"Bill, I don't know what to do." She needed to have her uterus considerably reduced in size so that it could be maneuvered into the body cavity, and a few sutures placed across the vulva to prevent a recurrence before it healed. She also needed to be on intravenous fluids and, very importantly, a heating pad to try to elevate her temperature. The cat was moribund

After a few minutes' discussion, I decided I would try to save her despite the circumstances, though I held out little hope for her survival. If my office had been unique in this blackout business, I could have referred him elsewhere, but the entire city was in the same fix. With the aid of the candles, we were able to see well enough to place an intravenous catheter into a vein so we could give her fluids. We perfused the swollen, protruding uterus from a bottle of the same solution that had been warmed in hot water. Then we proceeded to pour large amounts of granulated sugar all over the projecting organ. (Yes, sugar. The kind you put in your coffee, tea or cereal.) The sugar, absorbing water readily, draws fluid out of the engorged tissues, thereby reducing the size. As the first application of sugar became wet, we gently removed it and applied a new batch of dry sugar. When the size came down to a manageable diameter, we infused antibiotics over the surface of the organ and then bit by bit, very gently, manipulated it back into the body and placed a couple of sutures across her vulva.

So far, so good. We ran upstairs, wrapped a hot water bottle in a towel, and placed it with Samantha who was covered with a warm blanket. I came downstairs every half-hour to check her, turn her and see that the intravenous drip was still going. Despite all this, each time I checked her, her condition worsened. Samantha died seven hours after she came in. Though I was not surprised, it was discouraging

I telephoned Bill (at least the telephones were working) with the bad news. Bill was sorrowful and upset, but had expected this. The entire evening was one of eeriness: darkness, candles flickering, the shadows they threw, and this cat's misfortune under such conditions.

November 9, 1965 has always stuck in my mind.

Mixed Blessings

Sometimes you just can't win. I'm sure everyone has encountered situations wherein they deal with or correct a problem successfully, only to have the improvements lead to new, unforeseen difficulties.

Two cases come to mind.

Mrs. Kirkpatrick, an elderly, gray-haired lady, brought her 17-year-old cat in to see me. He was a grey and white male domestic short hair with a severe upper respiratory infection: his nostrils were solidly plugged with dried mucus, so that he had to breathe though his mouth, chest heaving, eyes full of pus, coat "unthrifty." He was extremely thin and dehydrated. Puss-in-Boots, as he was called, looked like he was about to cash in life number nine. It appeared he had an often-fatal viral respiratory disease called rhinotracheitis, so I was not too optimistic about his chances.

Nevertheless, we tried antibiotic injections, intravenous fluids, clearing ocular and nasal discharges regularly, as well as delivering nutrients by stomach tube.

Despite my extremely pessimistic prognosis, after a few days of treatment, this geriatric fellow looked like a new cat. He made amazing progress, and all of us involved in the treatment were extremely pleased. He not only ate, he begged for food each time someone passed by his kennel. His eyes were bright and clean, his nostrils clear, and he sneezed minimally. Even his coat was improved in this short time.

We sent him home with antibiotics and nutritional supplements. His owner, who had raised him from six weeks of age, was delighted.

A couple of weeks went by and Bill Hinkle, our veterinary technician, happened to wonder how he progressed at home.

We soon found out. Four weeks after Puss was released, we received the following letter, reproduced verbatim:

> Dear Dr. Wasserman,
>
> I truly appreciate everything your staff did for Puss-in-Boots. He felt so much better, that he attempted a leap which he never would have in recent months because of the shape he was in. He killed himself in the fall. It was very traumatic.
>
> Please accept that I do feel sincerely grateful for all you and your staff did.
>
> C. Kirkpatrick

* * * * *

Then there was Buffalo. Buffalo was a huge dog. He was almost the size of a small pony, and the sweetest, gentlest dog one will ever meet. I'm sure there was a fusion of several large breeds—some St. Bernard, a bit of Great Pyrenees, and possibly some Bull Mastiff hovering around in his genetic background. About six years old, he had an off-white color with a brown patch on the right side of his face. He had limpid brown eyes and a perpetually wagging tail, though at this moment it wagged not too enthusiastically.

"Doc," said his owner Steve as he entered the examining room, "Buffalo's belly has been getting larger and larger over a period of about a year. I thought he was just getting fat, but now he's sluggish and has quit eating. He hardly wants to get up. What's wrong?"

If Buffalo were a female dog, I would have guessed she was in the late stages of pregnancy and would soon deliver a large litter. In spite of his bulging abdomen, Buffalo was getting thinner, his spinal column and ribs were not too well covered, and he had a "hang-dog" look. It didn't take too much of an examination to determine that he had a monstrously large tumor in his abdomen.

I palpated Buffalo's abdomen and was sure the tumor involved his spleen. These tumors are sometimes benign, sometimes malignant. Surgery was required.

Laboratory work was done to ascertain whether he could withstand the surgery, and henceforth, we scheduled Buffalo for the operation. After anesthesia was underway and our technician Bill prepared the operative site, a long incision was made on the midline of Buffalo's abdomen, approximately 10 inches in length. My associate, Dr. Richard Turoff, and I, attempted to remove the mass through that opening, to no avail. We extended the incision a couple of inches anteriorly and posteriorly. We still could not get this monstrous growth out through this gaping wound.

We proceeded to the next step: two incisions starting from the middle of the existing one and extending perpendicularly to the right and left sides of the abdomen. Only then were we able to bring the growth out of the abdominal cavity, allowing us access and visibility so that we could tie and cut the numerous large blood vessels leading into and out of the spleen and remove this mass. What a job! Two hours later, we had him sutured up and in a kennel with an intravenous drip, heating pads, etc. The growth was about fifty percent larger than a basketball. It weighed in at 24 pounds, the largest we'd ever seen. The pathology laboratory confirmed it was a splenic hemangioma, which is benign but may recur.

In a few days Buffalo regained enough strength and appetite to return home. His owner paid part of the bill and promised to send the rest in installments.

Sutures were removed two weeks later. Buffalo looked healthy and happy, his tail wagging showing a decided vigor. He almost seemed grateful.

Several installments came on a monthly basis, and then in came Steve with the final payment. Of course I was anxious to hear how our spectacular case was doing.

"Steve, how is Buffalo faring? Okay?"

"Doc, I don't know. We don't have him anymore. He was feeling fantastic after his recovery from the surgery. Then about two months ago, a female dog in heat came by our house. Buffalo chased out after her and we haven't seen him since."

De Debbil

It was Wednesday, my day off from clinic hours, but I descended the stairs to my office to do some paperwork: pay bills, take inventory, order supplies, etc.

The telephone rang.

"Doctor, can you come over to Pier Ten? We've unloaded a horse from a ship and have been trying without success for two hours to get him into a van. Would you come over and give him a tranquilizer injection so we can handle him?"

The call came from the Red Hook piers, an area opposite the tip of Manhattan. On several occasions, I had been called aboard vessels from Sweden, Norway, Italy and other countries to examine dogs that had been picked up as mascots somewhere along their seagoing odysseys. I would give them rabies vaccines and issue health certificates. So, calls from the piers were no novelty. But this time, it was a call of another species.

Whew! This was a new one for me. I would just as soon pass this up, I thought. But, maybe . . .

"I want to make it clear," I said, "that I am not an equine practitioner and have very little experience with horses. Perhaps you can call around and get a veterinarian who does."

"We've called several, and the ones who have experience with horses are either unavailable today or don't want to travel over here. Please try to come." He sounded desperate. I told the caller that I'd get back to him shortly.

My scant experience treating horses passed through my mind. In veterinary school, I had once been assigned to treat a young Thoroughbred colt that had an enormous piece of flesh torn from his rump on a barbed wire fence. The treatment consisted of cleaning the area daily, scraping the edges of the wound to freshen them, and then literally painting the whole area with a red oily mixture, called appropriately, scarlet oil. This was a mixture of irritant compounds that stimulated new tissue growth, gradually filling in the deficient area. It worked remarkably well, but this experience would be of no help today.

On one occasion while I was treating this colt, he became extremely agitated and seemed to go berserk, kicking his hind legs wildly in all directions, causing me to scramble over the stall's wall to save myself from serious injury. It was scary, and I gained a respect for practicing caution around horses.

In the early days of my practice in Brooklyn, when there were still a few horse-drawn fruit and vegetable wagons plying their trade in the streets, I was sometimes called upon to treat one of these animals. On one such "house" call, I sutured a five-inch triangular flap of skin and muscle on the chest of a horse that'd had a losing encounter with a truck. This surgery was accomplished under the poorest of conditions, in a cold, damp garage, lit by a bare 40-watt bulb. It seemed almost impossible to prevent the site from becoming infected, let alone see what I was doing. Medicine is often helped along by the natural healing powers of the patient. Penicillin doesn't hurt either. Luckily, it turned out well.

Reminiscences over, I thought about the case at hand. I knew that a 50-pound dog would be nicely quieted down by a one-quarter of a ml. injection of the tranquilizer I used. So, if I dosed strictly by a ratio based on weight, a 1,000-pound horse should receive five ml. of the drug. Considering the vast diversity in species' reactions to drugs, I was very uneasy about calculating a dosage so simply. I couldn't take so foolhardy a chance.

Thus, I called a veterinarian I knew who practiced equine medicine at Belmont Race Track and described the problem. I learned the effective dose of this drug for a horse was only one and one half ml.! Five ml. could have killed the horse! Should I get involved with this horse call, I wondered? Why complicate my life? Yet it would add an interesting dimension to my strictly small animal practice.

I telephoned the horse owner and told him I'd be over soon. Placing the necessary supplies in my bag, I was off.

Picture this. I sauntered through the gate of the cyclone fence-enclosed pier, medical bag in hand. In the center of a large crowd were three men holding a large horse, one holding onto the animal's bridle straps and pulling, and two in the rear, holding the ends of a very wide belt against the animal's buttocks and pushing toward a ramp leading up to the door of a horse van. What a beautiful Thoroughbred stallion he was! Tall, he was 17½ hands high (a hand is four inches in horse circles), chestnut-colored, with a white blaze down the front of his face. He was perfectly proportioned, and had those beautiful dark brown equine eyes. He was a work of art. I loved gazing at horses, and especially watching them move. This one was not moving at all.

Almost simultaneously, all eyes turned toward the hero who had come to get the job done. I couldn't help but feel important despite my nervousness. The horse's owner, a tall, dark, well-dressed young man, approached and explained to me that he bought the two-year-old spirited Thoroughbred stallion in Ireland. Though the stallion had been led off the ship over two hours previously without a problem, nothing could persuade the beast to walk up the ramp into the van. The owner had a horse farm in Kentucky where they trained Thoroughbreds for racing. He felt that this expensive specimen had promise to become a great champion—but first, he had to get him home.

I approached the horse as though I did this sort of thing several times a day. Looking up at my wild-eyed patient, I laid my medical bag by my side and extracted a sterile syringe, the tranquilizer, and an alcohol-soaked cotton ball. I moistened a small area on the horse's chest with the antiseptic swab, drew up the recommended dose of tranquilizer and plunged the needle into the muscle, releasing the drug. No reaction from the horse. So far, so good.

The crowd had grown. At least 400 eyes were trained on us. Ten minutes passed. No change. Twenty minutes. The same, but an attempt was made to push-pull him up the ramp. No luck.

After 30 minutes, he began to sway from side to side, appearing drunk, as though he has imbibed in some of the Irish whiskey his homeland is famous for. The persuasion committee tried several more times to get him aboard the van, to no avail.

Finally, about an hour after the injection, with the full dose of the tranquilizer coursing through his arteries and nervous system, and in no small measure worn down by the persistence of his handlers, the three men pushed the horse, and to everyone's surprise, he began to move up the ramp haltingly. It was push, sway and move a few steps forward; push, sway and move a few steps forward. Soon he was in the van. Applause and shouts issued from the crowd, and a great sigh of relief issued from the veterinarian and from the owner.

The owner shook my hand and thanked me as I picked up my bag and started to depart. As I neared the gate, it occurred to me that this spirited creature might be worth following in the horse races.

"By the way," I called out to the owner, "what's his name?"

"De Debbil," was his reply. The Devil—good name.

When I arrived home, I realized that with all the excitement surrounding this unusual call, I had forgotten to ask for a fee for my services!

The Downside

Every veterinarian has downright upsetting, depressing, emotional episodes involving unexpected bad results from treatment or surgery on animals, including unanticipated deaths. Added to the stress on the veterinarian is the empathy he feels for the owner of a beloved pet.

Euthanizing patients that he may have treated for years may cause a good veterinarian to feel dejected, since it has severe ramifications on everyone involved. A few weeks ago, listening to the BBC, I heard that of all professions, the suicide rate is highest among veterinarians. The counselor/veterinarian conducting the study cited that the euthanasia of animals, often healthy animals, plays a large part in veterinarian depression. I was shocked, but in a way I understood.

In recent years, there have been considerable breakthroughs in veterinary medicine on ways to deal with the grief a dog or cat owner feels upon the demise of his or her pet. Grief counselors, generally social workers who are pet owners themselves, delved into the area of the special relationship of people with their pets. In New York City, The Animal Medical Center has 40 or more veterinarians on its staff and sees tens of thousands of patients each year. Bereavement counselors can be very important to grieving pet owners. People who do not own or even like animals would never understand the need for such help.

At local veterinary society meetings, occasional lectures are given for receptionists and technicians who work in small animal practices on dealing with pet loss. Lectures at veterinary conventions on the human/animal bond and grief over pet loss are more common now than ever before.

However, most veterinarians instinctively knew how to deal with a client whose pet died, well before the "discovery" that people mourned deeply at such a time. How do we know how to react? Because we generally love animals, and have lost our own pets, so we know how a client feels. It is as though the bell tolls for our own pets. I personally have known only two veterinarians who failed in practice. They had no pets themselves, and generally felt that owners did not value their animals enough to spend money on diagnostic tests, x-rays or major surgical procedures, so they did not offer full services.

Following are two examples of the "downers" we experience.

* * * * *

The Flemings had two Afghan hounds. Both were about eight years of age, tall and thin with honey-colored hair and black markings on their muzzles and the tips of their ears, lucid brown eyes and sweet dispositions. Khan, the male, was large-boned and seemed to have more medical problems, so he was in more often than the daintier and extremely shy Shamal, the female. She was extremely fearful of all but her owners, which was the main reason she wasn't brought in too frequently.

At the very beginning of my practice, we offered a small amount of bathing and grooming of dogs. One morning, Mrs. Fleming left a very matted and frightened Shamal for grooming and bathing. The dog shook and cringed at the back of her kennel, but everything seemed to go well while she was being groomed, and afterward as well.

That evening, the Flemings came to pick her up. As they walked her out the door to where their auto was parked at the curb, Shamal collapsed. Her owner picked her up and rushed back in and I cleared the client and patient from my examination room for this emergency. Within seconds of being placed on the table, it was obvious her life was ebbing and Shamal died.

We all were stunned and appalled! How could this have happened? She hadn't even come in for a medical problem, only a bath and brushing! I felt as though I had been hit in the midriff by a sledgehammer, knowing both the anguish these gentle people felt and my own feeling of grief and disbelief at seeing this sweet animal cross unexpectedly from life to death in a matter of minutes for no discernible reason.

I asked the Flemings if they could wait a short while so that I could finish office hours and then perform a postmortem examination. Though distraught and tearful, they agreed. Their pets were children to them, and like them, I was shocked and distressed by this disaster and like them, I could not have rested unless I found out what had occurred.

Immediately upon incising the abdominal wall, it was apparent that the entire abdomen was filled with blood. We sponged out the blood as best we could until we could see and feel the abdominal Aorta, the main and largest artery in the body. Tracing it posteriorly, we came upon a "blowout"—an area of the artery wall that had thinned and formed a bubble (somewhat like an old tire inner tube) that had ruptured. We call the condition an aortic aneurysm. It had existed for a long time without causing any problem, and under the stress of being left with strangers, the aneurysm had ruptured. This sometimes happens in humans as well.

So poor Shamal was gone and we were all wrenched by

her passing, but we knew nothing could have saved her; it was not foreseen. The Flemings left tearfully, taking their pet with them for burial. I hoped neither they nor I would ever experience anything like that again. On a happier note, Khan lived for several more years, though undoubtedly missing his companion, as did the Flemings, I'm sure.

* * * * *

The most outstanding case in my memory is that of the Eismanns and their dog. Bernard and Suzanne Eismann had a large, beautiful nine-year-old black-and-tan German Shepherd who had been a patient for several years. One could see the love and attachment the Eismanns had for their dog by the warm smile on their faces whenever they looked at or talked about Grendel. The Eismanns had a country home in the Berkshire Mountains in Massachusetts where they took their dog on weekends, holidays and vacations.

On one such visit, Grendel, roaming free in the country, had not returned by the time the family was ready to start their trip back to the city. Bernie postponed his own departure, sending Suzanne and their two children home to work and school. He scoured the countryside, asking everyone he saw if they had seen his dog. Heartsick, he returned to the city after a few days. He could think of nothing beside the loss of his dog—all the possible disasters that could befall her, and the fear and bewilderment that his canine friend must be experiencing. As a writer and correspondent for Public Television, he found it impossible to concentrate on his work.

Bernie returned to the mountains a few days later and continued his search, trekking through forests, calling out his dog's name and then widening the parameters of his reconnaissance by automobile. He posted notice after notice with Grendel's description and an offer of reward money—with no success.

He followed this routine for three weeks, and, depressed and about to return to the city (but not to abandon the search), he was sitting on his porch towards twilight, and who came walking up the path, slowly, haltingly, lame, barely able to propel herself forward, but Grendel!! Bernie described this meeting: he cried, he got down on the floor and embraced and kissed his friend, who yelped with joy, showering Bernie's face with licks. They rolled around on the porch floor. They couldn't get close enough to each other. Love? Believe it. This went on for some time and then Bernie realized some food and water would be in order.

The following morning, Grendel was in my office for an examination. Dry, flaky, matted with scattered sores and numerous cuts and bruises, her coat and skin were in terrible condition. And she was emaciated. Her eyes were sunk into their sockets, and her skull bones, spinal column and hip bones showed prominently. Yet basically, she appeared healthy—nothing that a few weeks of good nutrition, grooming, and love wouldn't cure.

Where had she been? What did she subsist on for three weeks? Did she revert back to her feral ancestry and hunt small animals? Had someone captured, imprisoned and mistreated her? Unfortunately we will never know.

Grendel recovered and gained back her lost weight, but age took its toll on her. About two years after her mysterious absence, like most Shepherds who reach 11 or 12 years of age, her hips deteriorated, developing arthritis, accompanied by neurological degeneration. She could no longer walk or stand, even to eliminate wastes. It was as distressing for her owners to observe as it was for Grendel to bear.

Suzanne and Bernie brought Grendel in for euthanasia. Bernie literally carried the dog and asked if he could be present while it was done. Tears streamed down his face as he spoke gently to his friend, embracing her around the neck, his face on Grendel's, saying goodbye.

I shed a few quiet tears myself up to this point, but after it was done, the floodgates really opened. Bernie sobbed convulsively, cursed, and fell on his dog, hugging and kissing her. And then, he and I looked at each other, and I, seeing and feeling his grief, started sobbing unabashedly as well, and we embraced—two crying men standing there in the examining room, holding on to each other for several minutes. Suzanne, outside the examining room, poured out her grief as well.

When they had calmed down sufficiently, he and Suzanne said goodbye, taking their friend with them for burial in the country.

Recalling that day as I write this, I can feel a lump forming in my throat, and tears almost shed.

Grendel, with Suzanne and friend.

Above: Grendel, age 2, with Mariann and Jonathan Eismann.
Below: As a pup, with Mariann.

Alert

The loudspeakers blared instructions with great urgency from the street. It was 3 a.m. on the coldest February I could remember.

"Everybody on this street get dressed immediately. Be prepared to evacuate your homes." This was repeated over and over and at first, I thought it was part of a dream. Rubbing the sleep from our eyes, my wife Bernice and I went to the bedroom window. The scene outside revealed police cars and gathering groups of people, the clang of fire engine bells, and the acrid smell of smoke. I put on a robe and went upstairs to the room where my sons Harvey and Andrew slept.

From the deck outside their room, I noticed clumps of flaming debris flying by the house, some appearing to be the size of automobiles, driven by the wind. It looked frightening and surreal, like something seen only in movies, concocted by special effects experts. Above the surrounding buildings two blocks southwest of my house, I saw flames shooting high into the sky. I knew then that the newly renovated 13-story Margaret Hotel was being demolished by fire. Luckily, I thought, it was not yet occupied. I had never experienced anything like this, especially the fiery flying missiles, which will be forever burned in my memory.

The reason the fire could so easily be seen, and the debris fly so freely through the streets, was that most of the buildings were two and three stories high and the blazing Margaret, at 13 stories, towered over the neighborhood. I was mesmerized, and although there was a tendency to stay and watch, I knew the danger was real, and I had much to do.

Our 1822 federal-style building was mostly of wood construction, as were many other buildings in the area. One landing clump of burning debris would mean rapid ignition and incineration. Dressing frantically, Bernice and I made two piles of warm clothing and told the boys to dress quickly and be ready to leave—pronto.

"Find a couple of shopping bags. We may be able to take some valuables if you can lay your hands them on quickly, I suggested to Bernice. "Are the kids ready?"

Just then, they emerged from the stairway, sleepy-eyed and with questioning looks.

"Boys, the Margaret hotel is burning and we may have to leave the house. Get the cat carriers from the closet and put Benjie and Homerella in them." Benjie, a large, gentle, intelligent, male seal-point Siamese, and Homerella, a sweet, petite female seal-point Siamese, both treasured pets, had to be pried out from under a bed where they hid in fright.

I thought about trying to look for my passport, birth certificate, other important papers, mementos and pictures, but decided it would take a lot of rummaging and there wasn't time. Who is prepared for a situation like this?

We all walked down the flight of stairs leading to the front door and gathered together to await further instructions. My sons each clutched a favorite possession to his chest and the cats meowed in their carriers.

So far so good. We had come this far without mishap. At least our lives would be saved.

"Oh my God," I suddenly cried in panic. As I reached the office door, I realized I was far from ready to walk out. There were five hospitalized patients in the clinic, all cats, two of them receiving intravenous fluids. They needed to be readied for evacuation too.

I thought, "What a ghastly situation. They don't teach you anything about this in veterinary school."

"Come on everyone, I need help," I shouted, with some panic evident in my voice. I opened the door to my office and my family immediately understood and followed.

Once the patients were disengaged from intravenous drips, we carefully placed them into carriers left by their owners. All were wrapped in heavy towels to ward off the cold. Now the waiting contingent consisted of four humans, serenaded by a chorus of seven cats, with musical counterpoint provided by loudspeakers, sirens, gongs, and the buzz of conversation of the ever-increasing crowds.

I walked outside. The flaming missiles still flew by, but fewer in number and size. The torrential water dousing the burning building three blocks away was driven by the wind to all the neighboring streets. Water coated trees, buildings, fire hydrants, and automobiles, freezing instantly and turning to ice. Once frozen, the trees appeared to be made of a delicate, fine crystal. It was a beautiful sight, but due to the situation at hand, could not be enjoyed in a relaxed, lingering way. My family's anxiety was almost palpable. Time seemed frozen, like the trees outside, but it was actually 5:30 a.m. and the first light of dawn illuminated the horizon.

Suddenly, the loudspeaker message changed.

"You may all return to your homes. The danger is over. Repeat. Return to your homes. The danger is over."

Weary but relieved, we returned the pets to the safety and warmth of their kennels, re-established the intravenous drips, and trod slowly up the stairs with the cat carriers containing Benjie and Homerella, who were still complaining, and our other assorted packages and objects.

Harvey and Andy went to their beds once we assured

them that they were safe and should get some sleep. Andy drifted right off.

But there was no sleep for my son Harvey, a tousle-haired entrepreneurial type. He had gotten himself out of bed, dressed, and, grabbing his camera, ran out of the house and took pictures all over the neighborhood: of the fire, of firemen still playing streams of water at the building, of the crystalline trees and the onlookers. He then called The New York Times and asked if they were interested in the pictures. They were. Traveling by subway, he dropped the rolls of film off at the newspaper. Returning home and obviously deprived of sleep, he nevertheless waited for a call from the newspaper. When it came, it was most disappointing. The pictures were unusable—the shutter speed had not been right for that low level of light.

The idea that danger, even death, could come suddenly in the night—was that over? The chance of losing my building and practice, my family and pets, as well as other people's' beloved animals? The specter that all this could vanish suddenly—was that buried? I knew my casual acceptance of good fortune and safety had forever changed in these few hours. And I was thankful, and hoped never to forget what was important in life.

Oil Spill

It was mid-February and I arranged to take a week off to ski with my 12-year-old son Andrew. Andy and I planned to go on several one or two-day outings to areas within two to three hours of New York City.

The days were quite warm during the preceding week, and on the day of our outing, the temperature must have reached a record high. The morning started warm, bright and sunny with a temperature of 45 degrees Fahrenheit that reached the mid-50's by early afternoon. Whatever snow remained on the mountains turned into streams of water sluicing down the slopes. Skiing was out, I realized with great disappointment.

It hadn't been easy to get a locum tenens (as the English refer to a temporary fill-in doctor for a solo practice). I had to find a freelance vet, not in his own practice or working for someone as an associate; I wanted someone capable who would represent my practice well and who was free at just that time. These qualities had all come together in my temp, and the weather was more conducive to swimming than skiing, thus spoiling my careful plans.

And then, listening to a radio news broadcast that morning, I learned an oil barge had struck a rock in the Hudson River near Bear Mountain, New York, and spilled 400,000 gallons of oil into the river ten days earlier. A cleanup by the Coast Guard was in progress. Headquarters for the cleaning and treatment of waterfowl had just been set up the previous day in a Rockland County highway garage.

Being a bird enthusiast, I called the radio station to get the telephone number and location of the garage. Stony Point, New York was about an hour and a half from Brooklyn. The headquarters readily accepted my offer of volunteer professional help, along with that of my 12-year-old assistant. Andrew, a bright, inquisitive animal lover, was eager to come along. The rescue people suggested some medications and supplies they could use. We gathered them from the office and departed.

What a sight greeted us! The Rockland County highway garage was an unpartitioned, multi-windowed structure about a city block long, half as wide and at least three or four stories high. The Coast Guard and local humane organization had attracted an army of volunteers: young people on break from school and eager to help, elderly retired folk, wanting to contribute, and all ages in between. One older couple from upstate said they closed their store so they could come and help. There were more volunteers than tasks, and the overflow crowded around the wide doors, watching the activities inside.

Since I was on vacation, I was the only full-time veterinarian working there; other local veterinarians came during their lunch hours, between office hours, after surgery, and spent two or three hours at a stretch. The head of the local humane society and animal shelter was in charge of organizing the bird treatment and cleanup.

I had never treated birds subjected to an oil spill, but I learned quickly with the aid of the Humane Society people and a couple of the local vets who had worked there the previous day.

Oil-covered birds brought in by the Coast Guard and volunteers were evaluated for condition by a veterinarian. If they were in a state of shock and poor condition, the birds were treated and allowed to rest in pens, covered with towels and rags, and heat lamps were directed at them, before they underwent the washing procedure. Otherwise, they were washed at once. Low body temperature (106 Fahrenheit is normal for birds!), depression, shallow breathing, and loss of balance were criteria for treatment. There were a few score birds in various stages of treatment.

Treatment consisted of an intramuscular injection of Azium, a form of cortisone, to counteract shock; milk of magnesia given by stomach tube helped evacuate ingested oil, largely gotten by preening; a dextrose and electrolyte solution (in this case Gatorade was used) was also administered by stomach tube. Then we coated the birds' eyes with ointment for protection during the washing process, and their beaks were taped to prevent them from preening and ingesting more oil.

I learned to do these things quickly and with ease. At one point I went into a pen with eight swans and taped their beaks with masking tape. I'd heard that swans could be vicious, but somehow I had a feeling they would not resist, and they didn't. They knew they were being helped.

Beside swans, there were Canada geese, ducks (mergansers, buffle-heads, mallards) and some gulls. Only three birds died on the previous day.

An assembly line was set up for washing. Large galvanized tubs lined a long series of tables laid end to end. The first four or five tubs contained warm Lux liquid cleaning solution, the next four contained warm water rinses. Two eager volunteers at each tub wore gloves and coveralls or aprons. Sleeves rolled up, one person held, one washed.

I knew nothing about fuel oil. Oil was used for autos, for heating and to operate generators. Oil was oil. But the substance in this spill was an oil of a different oleaginous stripe. It was #6 fuel oil, also known as Bunker-C, a very heavy concoction—almost a sibling of tar.

The viscous oil proved too much for the mild Lux. After a telephone call was made to the West Coast where spills are more common and experience in treating victimized waterfowl is greater, several drums of Shell Sol-70 were dispatched immediately.

Andy and I drove back to Brooklyn that evening feeling very tired but fulfilled. He had been holding birds while I administered treatment via stomach tube and injection.

When we returned the next morning, the new cleaner was already there, flown in by courtesy of Shell Oil Company. I thought it was only right that oil companies should contribute to cleaning up the destruction they cause. Shell Sol-70 proved far more effective in removing the tarry substance.

A petroleum solvent, Shell Sol-70 is highly flammable and gives off noxious fumes. Great care must be exercised. Because the solvent dissolves plastic, washers were issued sturdy rubber gloves and appropriate clothing. Washers were rotated regularly among the abundant volunteers to avoid skin rashes and eye damage. Exhaust fans were placed to dissipate the fumes from the building. The birds' heads could not be immersed but were carefully washed with a solvent-soaked cloth, then wiped until the oil was gone. After immersing the body and wings to remove much of the oil, the birds were passed along and the process repeated in three more tubs, tarry solutions being replaced after every five birds.

After a wash, the birds were depressed and uncoordinated. For several hours, they could only lie down or stagger, taking slow, deep respirations. The treatment nearly killed them until the solvent evaporated. We administered fluids and anti-shock injections twice and three times to a number of birds. The grain-eaters that had not eaten for two days after washing were tube-fed a mixture of chicken mash and Karo syrup mixed in a blender, and fish eaters were fed pureed fish. Birds showing signs of infection were given antibiotic injections.

After surviving all these drastic ministrations, the poor birds were not yet out of death's way. Luckily, experts in shore bird ornithology orchestrated the next step of the rescue. The solvent apparently removed the oil from the birds' skin and feathers so that they were no longer waterproof. The birds were put into one of a number of children's swimming pools and taken in and out of the water periodically for six to eight hours. They had to be watched carefully because until they produced their own oils again, they would sink. The pool regime continued until the birds preened and showed they could stay afloat.

A few birds ingested a great deal of oil as evidenced by blood in their stool, by tremors and other neurological symptoms. They looked pitiful as they were convulsing—a heartrending, nauseating sight. I was told it was futile to treat these birds. They died shortly, gasping out their last breaths.

Late in the afternoon of the second day, no new birds came in and those present received all the treatment that the veterinarians and other volunteers could give. Andy and I were saying our goodbyes, when suddenly, a joyous chorus of shouting and applause arose from those assembled in the building. We looked up and saw a flight of Canadian geese, 15 or more, circling around and honking in the interior of the enormous structure! Imagine seeing this sight indoors! Spontaneous applause broke out. What a great reward. It seemed like a thank-you flight.

We drove home weary in the twilight, and knew we would remember this wonderful and fulfilling experience.

Tuxedo

Tuxedo was an aptly dubbed feline. All black, with a white bib and some white on all four paws, Tuxedo looked like he was on his way to a swanky dinner party. The large, very gentle male cat with beautiful pale yellow eyes was brought into my clinic because of urinary tract blockage, part of a condition called Feline Urological Syndrome.

FUS involves the production of minuscule crystals in the cat's urinary bladder. The inability to pass these out through the urethra creates a blockage. Urine backs up and severely damages the kidneys, causing uremic poisoning and death. Male cats are primary candidates for FUS, and when blockage occurs several times despite treatment, a surgical procedure called perineal urethrostomy is recommended. It involves creating a larger urethral opening in the same area.

Such was Tuxedo's case. He was brought in at age five with his third blockage. His bladder should have been grape-sized. Instead it was the size of an orange and rock-hard. He was vomiting, deathly ill with uremic poisoning.

I discussed the operation with Tuxedo's owner and recommended that it be done after detoxifying his pet. I told him what the surgery and hospitalization would cost. He patted Tuxedo on the head and said, "You're a nice fellow but you're not worth that much." Even though I knew he could easily pay for it all at once, I offered to let him pay it over a period of time but he insisted that his pet be put to sleep. I couldn't change his mind. He paid for euthanasia and left, not even pausing to say good-bye to his friend.

Well, I couldn't kill this gentle, sweet creature just because his owner was an insensitive, cold, uncaring fool. We relieved the blockage and gave Tuxedo intravenous fluids to detoxify and prepare him for his surgery a few days later. The patient made an uneventful recovery. He became our clinic cat—a lovable rascal.

Between office hours, Tuxedo roamed free in the clinic and spent most of his time trying to mooch more food. On one occasion, I passed by a kennel housing a particularly vicious cat that lunged through the bars and nearly caught my arm with his steely claws. Tuxedo, the mild, sweet pussycat, was nearby emitting wild ferocious growls. He leapt up to attack the offender who then backed off. It was almost as though Tuxedo said, "You've done well by me and now I will protect you."

We had Tuxedo for about two months and decided to convince my mother-in-law to take him. She was 83 years of age and lived alone. Her previous pet, a Siamese cat named Miss Kitty, died three years earlier. She was very distraught over her pet's death and had refused to get another cat.

One day, we just showed up with Tuxedo. Mild protests gave way to a conditional acceptance. We'll see how it works out. It didn't take too long for her to get hooked on Tuxedo, and since he asked for food constantly, she obliged him. In no time, the slim, lithe animal looked like an overstuffed black-and-white cushion. He became enormous, weighing in at 20 pounds.

Months later, my mother-in-law fell and broke her hip. She was due for a long period of recuperation, and it was dangerous for her to have Tuxedo underfoot where he was most of the time, begging for food. It soon became evident that she would not be able to care for him again. I took him back to the hospital and gave strict orders about very minimal feeding.

Within a few weeks, Tuxedo was back to normal cat size. Shortly thereafter, a young couple, Mr. and Mrs. Martin Fishman, came in to ask where they could adopt a nice cat. What a question!

I saw him eight years later, and though aging, he was in pretty good shape. The Fishmans were crazy about him. He's still got several lives left, I thought.

Very recently, the technician in the practice phoned me upstairs to tell me the Fishmans were in with Tuxedo. Did I want to see him? Did I!

The Fishmans were in the waiting room when I came down, with a very thin Tuxedo. Tuxedo was wasting away from the common nemesis of aging cats: kidney deterioration, always a losing proposition. His face looked shrunken: his eyes sunk deep in their sockets, his lusterless skin hung loose. I lifted the delicate Tuxedo and took him upstairs for Bernice to see. She held and hugged him and we reflected on how many people had gotten pleasure from him. By his appearance, I knew he would not live much longer.

Two weeks later, Bill let me know the cat was downstairs. This time he was all but dead. His only signs of life were his eyes and his faint breathing movements. He was in to be euthanized. Mrs. Fishman cried outside. They dreaded telling their young daughter what was to be done.

Tuxedo was 17 years old.

Tuxedo's first owner still lives in the neighborhood and we pass in the street occasionally. To this day, he is totally unaware that his pet had a 12-year "life after death" experience.

Above: Tuxedo.
Below left:With the Fishman's daughter.
Below right: With Martin Fishman.

Beginnings

How did my career in veterinary medicine start?

The year was 1940. The New York Times was three cents, a subway ride was a nickel and the cost of a hot dog was five cents. My friend Irving Lieberman and I sat in the lunchroom at Brooklyn College, commiserating with each other. His elderly dog just died, and mine, a white and brown male terrier-spitz cross had been "put to sleep" after suffering a stroke which paralyzed one side of his body. It was a sad time. We were both third-year students considering possibilities for a career and had visited practitioners of several professions. We had honed in on veterinary medicine as our choice.

Between classes, we often walked the railroad tracks behind the campus, discussing our plans and deciding which vet schools might afford us the best chance of acceptance.

During the following year, Irv and I hitchhiked up to Ithaca, New York to visit the veterinary college at Cornell University. Our mothers packed numerous sandwiches which we ate in wheat and hay fields alongside the highway between rides. I don't remember where we stayed, but we probably had enough money for a couple of nights' lodging.

At Cornell we were impressed, excited, and awed by what we saw. We watched surgery performed on a horse and a cow, and the treatment of dogs and cats in the small animal clinic. We decided right then and there that we wanted to be a part of this profession.

The dean of the Veterinary College was not very encouraging. He explained to us that for each entering class of 60 students, Cornell received many hundreds of applications, and generally they preferred people with farm experience. It didn't look good for two young men from the city.

Irv and I made a similar excursion to the only other veterinary school in the region, the University of Pennsylvania in Philadelphia. It was a lovely old school with a cobblestone courtyard, I remember. Unfortunately, the response was the same. At that time, the 10 or 11 vet schools in the country each took 60 students a year. Such schools were funded by the federal and state governments and were obliged to give preference to residents of their state and neighboring states without a vet school of their own.

Nevertheless, we applied to several schools.

By our graduation in June of 1941, we had struck out in all our choices. Irv had also applied to a new school in Massachusetts, which was not yet recognized for licensure of their graduates. It was hoped that the school would soon be accredited, but it turned out most graduates had to go an additional year to another vet school to qualify to take licensing exams. Irv started there that fall.

Though jobs were scarce, I managed to pass a civil service examination for a clerical job in Washington, DC. I remember the annual salary was $1,080! Can you imagine that? Everything was correspondingly low. I lived in a boarding house for about $10 a week with meals.

Six months after my residence in Washington, the attack on Pearl Harbor occurred and one month later, I was in the U.S. Army. Bye-bye vet school for awhile. Though I felt frustrated, I knew I wanted to serve in the army and that I would try to enter vet school again. The army discharged me after four years and I applied again, holding various nondescript jobs in the meantime.

I waited, hopeful but realistic, for some good news.

And then . . . and then . . . in the spring of 1946, I received an acceptance letter from the veterinary school at Michigan State University. Hallelujah! I was elated at the prospect of school in the fall.

But the bell tolled for me in the form of a letter from the veterinary school, arriving about a month before I was to report. The letter said that I was not accepted, that there had been an error. I was stunned. Letters flew back and forth, and I finally paid a visit to the college, speaking to President John A. Hannah and a number of pertinent Veterinary College staff. I learned unofficially that an influential state legislator had insisted that the school admit the son of a constituent. As a non-resident from a distant state, I was the logical one to be bumped. Sometime during the conversation, I was unofficially given to understand that if I were to stay in Michigan, work, establish residency and perhaps take a few courses and reapply, I might be admitted the following fall.

I took a job working the night shift at the Fisher bodies plant in Lansing, a division of General Motors Corporation that assembled bodies for one of their models, a convertible. I worked on an assembly line from 4 p.m. to midnight, jumping into a convertible, tightening four bolts using a jig and an electric wrench, jumping out and into the next convertible, repeating the operation, and so on. I used to sing at the top of my voice while working. No one heard me, not even the men working a few feet away. I imagine deafness is a common ailment of men who work in such plants for many years.

Despite overalls, my arms and legs were blackened with grease every night. The company sometimes gradually increased the speed of the assembly line in order to turn out more cars per hour, hoping the workers wouldn't notice. Instead, the workers had to run to get the job done in the allotted space. Many a bolt was never bolted or a weld welded when this happened. Talk about a Charlie Chaplin operation!

I enrolled in a couple of microbiology courses during the day. The two A's I earned didn't hurt. When I reapplied to the vet school I was accepted for the following fall—this time, without error. And as those telling, terminal words in Philip Roth's novel Portnoy's Complaint read, "Now ve begin."

* * * * *

My first year of veterinary school was grueling, almost to the point of abandonment. Not only did I need to learn and memorize the names and locations of all the bones, their protuberances, grooves and openings, but their functions as well. Then the muscular system and the seemingly thousands of muscles, with their locations, origins and insertions, tendons and ligaments and their modes of action were next on my list. Then came the heart and circulatory system, arteries and veins with all their branches (bifurcations as they were called), and the organs they served; the brain and spinal cord and myriad of nerves and their branches, as well as actions they had on organs of the body when stimulated; the integument (skin, hooves, horns, etc.); the gastro-intestinal organs; the urinary system; the reproductive system—these were among the many things I had to learn. Not only was all this information incredibly voluminous, but I had to know it for about six species of animals! Oh, yes—I almost forgot about chickens, which have their own unique anatomy and seemingly hundreds of diseases.

Histology, the microscopic anatomy of tissues, and

embryology, the study of the developing fetus, were fascinating, and held my interest during the second year.

During the third year, besides lectures on the diseases of all species of domestic and companion animals, we spent time in the large and small animal clinics, watching treatments and surgeries performed on dogs and cats, horses, cattle and sheep. Unfortunately, our hands-on participation was minimal in those days. We also went out on farm calls with the professors of large animal medicine. This was fascinating to me, a city fellow.

* * * * *

Sometime during the third year, I went for a haircut at my usual barbershop. The barber engaged me in conversation, and on learning that I was a veterinary student, asked if I would spay his dog. I told him that although I had watched a number of spays (ovario-hysterectomies, done to prevent pregnancies and breast tumors), I had never done one. He said he'd be willing to let me if I wanted to.

Did I want to? Does a child want ice cream? I was excited and very frightened at the same time. What if something went wrong and I killed his dog? Still, I couldn't resist the chance of operating on my own, and so I agreed. There was no question of a fee, of course. I believe I would have paid him for the experience.

I gathered all the materials I needed for the surgery: anesthetic, syringes, instruments, suture material, needles, antiseptic solution and ties to secure the dog to the table.

The barber drove me out to his rural home. My patient was a female English setter named Susie, about eight months old, white, with brown flecks over her body. She had a sweet and happy disposition, causing my heart to sink in my chest. What if she died? I couldn't stand that thought, and I grappled with the urge to back out. But her owner had more confidence than I did; perhaps he had been through this before with other students, other dogs. We prepared for the spay on his kitchen table, with a standing living room lamp next to it. Ready or not, I was going to be a surgeon.

Only someone who has completed a major surgical procedure for the first time, alone, and under primitive conditions could know the fear and trepidation I felt throughout the course of the operation. I was soaking wet with perspiration when the last suture was placed. But when Susie started to yelp as she was coming out of anesthesia, I felt indescribably elated. I had done it, and the patient was all right!

The barber took it all in stride and insisted I take home a large chunk of venison he had in his freezer. I tried to refuse, not liking the idea of eating deer. I gave it to my friends in Lansing, Norman and Nancy Berkowitz. Nancy prepared it and invited me to dinner. The meat was gamey and we couldn't eat it. We were unaware that game should be marinated first.

I did one more spay after that, also on a kitchen table, for a friend who lived in Detroit. It was a cocker spaniel. I sweated it out, but a tiny bit less this time. After all, I was an experienced surgeon!

* * * * *

The summer between my third and fourth year, I was accepted into a program working with a government veterinarian who drew blood from cattle to test for a disease called brucellosis. Cows with this disease would abort their calves, causing serious problems for both dairy and beef cattle farmers. It is very contagious, and can also be transmitted to humans. It is called undulant fever in man, and is difficult, if not impossible, to cure. Affected cattle are destroyed. People who get it are treated intermittently throughout their lives when they develop high fevers.

I was assigned to work with a veterinarian in northwestern Wisconsin, an area called the Indian Head country. (I never really found out why it was called that, but vaguely recall someone saying it had to do with the geographic outline of the area.). The vet, a large man in all dimensions, was Dr. Raymond Pinkert, but was known by friends as "Tiny." Tiny Pinkert was one of the gentlest, nicest people I have ever met. A blue-eyed, shy, soft-spoken man in his early forties, he was originally Canadian but had settled in Wisconsin many years earlier. There was no guile or pretense about Dr. Pinkert. He was the genuine article.

We started our day at 5:30 a.m. or 6 a.m. from Amery, Wisconsin—population about 560—and drew blood from two or more herds of dairy cows that we lined up the previous day. We instructed the farmer to leave the cows in the barn after milking, for our access. Occasionally we found a few calves straggling out in the muddy yard, and we'd have to wrestle them in order to draw the blood samples. In one instance, during a tough session, I stopped and wondered what a city boy like me was doing in Wisconsin, up to my knees in mud and cow manure and wrestling with calves. It seemed so incongruous, but it was a very satisfying adventure. The hard-won samples were sent to a government laboratory in Madison. We tested for tuberculosis as well, especially on suspect cows.

On weekends, Tiny went home to his wife and young son in another small town in Wisconsin. I got together with a classmate of mine, Bill Walquist, who was assigned to another vet in the area. We fished in local streams, or watched baseball games played by itinerant semi-pro teams, or hitchhiked to Minneapolis. We also managed to do a lot of reading.

Toward the end of the summer, Tiny and I started to draw two days' worth of blood samples in one day so we could take the next day off. Very early in the morning, about 6 a.m., we rented a boat on one of the many lakes in the region and went fishing. The peace and quiet were almost palpable. I saw my first loons, with their crazy cries and diving antics. I never knew where they were going to emerge when they dove. Tiny filleted the sunfish and perch that we had caught, fried them in butter at the ever-present grills in picnic areas, and, accompanied by corn-on-the-cob, and salad vegetables that we had bought, we feasted in the fragrant, forested outdoors.

As this idyllic summer came to a close, I felt fortified to tackle my last year at school. Regretfully, I did not keep in touch with someone who was important to me, and years later when I tried to look Tiny up in the directory of veterinarians, he was not listed. It has remained a great regret of mine that I didn't maintain a link with him.

Recently while perusing a directory, which lists most of the veterinarians in the United States, I came across the name Dr. Paul Pinkert who lived in Madison, Wisconsin. Hoping he was related to my Dr. Pinkert, I wrote to him and received a gracious letter in return. No, he was not related, but he had met Raymond Pinkert several years earlier. They had probed for a possible familial tie but there wasn't any. He had not seen or heard about Raymond since. I then wrote, saying that I would appreciate it if, without too much trouble, and providing that Raymond's son still lived in Wisconsin, he could locate his whereabouts. I think a son would appreciate hearing what a colleague thought about his father. I haven't heard back so far.

My friend Irving, who was a brilliant student, graduated from the vet school at Brandeis University and went to work for the United Nations Relief and Rehabilitation

Agency accompanying shiploads of cattle and horses to devastated countries in Europe after World War II. He later ended up in teaching and research at a couple of medical schools. Due to neglect on my part during a particularly difficult time in my life, I lost contact with him also.

Smugglers

It all started when a very good British friend who lived in the United States for many years found it necessary to move back to England. But she had a complication—her two beloved cats. They were 17 and 18, and they had lived with her since they were kittens. Why was this a problem? Why not ship them by air as many thousands of animals are shipped all over the world every year?

England is one of five or six countries that have eradicated rabies, a viral infection that affects all mammals, including man. Rabies is transmitted by the bite of an infected animal or by eating the flesh of one, as it is spread in the wilderness. Domestic pets would have to be exposed to and bitten by an infected animal. The symptoms are primarily neurological and horrendous, and the victim's infection usually ends in death.

In order to keep its island-nation free of this disease, Great Britain has a six-month quarantine period for all dogs and cats entering the country, since the disease manifests itself symptomatically within that period after contact with a source of infection. So, for six months, a dog or cat entering the country must stay in a kennel under veterinary supervision.

Seventeen- and eighteen-year-old cats generally do not have much of a life span left, and six months away from their owner could easily hasten their demise. Since these cats lived almost all of their lives as indoor cats, had never been bitten, and, for good measure, were routinely vaccinated to prevent rabies, there was no chance they could carry the rabies virus. Marjorie knew of no one who would take on her geriatric pets, so, unless she could avoid quarantine, the only other choice was euthanasia.

So, what to do? My friend Marjorie said that under no circumstances would she have her pets killed. She asked me for some pet tranquilizers, and dosing instructions for April, a sweet female calico whose life had spanned 17 Aprils, and Sinbad, a large, black, friendly and talkative 18-year-old male. I informed her of the possible side effects and the approximate onset and duration of action. She told me nothing about her plans, and frankly, I really didn't want to know. I found out soon enough.

She arranged to leave her cats at a pet shop that boarded animals, operated by a friend of mine, Bernie Gordon. Bernie was to supply shipping crates for the cats and, when given the word to do so, deliver them to an airline that would fly them to Paris, France.

Marjorie had flown back to her home in England, a beautiful area consisting of 13th-century towns known as the Cotswolds. She planned to drive down to London to pick up her French-speaking friend Peggy, and then drive to the south coast. From there, she and Peggy would take the ferry across the channel to France, and drive to Paris to be there when the cats arrived. She would then tranquilize them, hide them in the car, and smuggle them across the channel into England on a late-night ferry.

Years earlier, Peggy had done this with her own pet, a miniature Dachshund, so Marjorie felt somewhat reassured of a chance of success for this venture.

Before leaving home she looked for the tranquilizer drops I gave her. They were nowhere to be found—a bad start. Peggy mentioned that she had some tranquilizer pills left over from her own smuggling foray, but she couldn't find hers either. More bad luck. They left for the coast in Marjorie's very small car, a Morris 1100, to get the ferry over to France. Once there, they drove three hours to a small country village about an hour from Paris, where Peggy had some friends who helped the two plotters find a place where they could stay with the cats for a few days when they arrived from the United States.

They began to feel discouraged about the plan. They had no medicines, and there were probably numerous unforeseeable problems and pitfalls ahead.

On the Sunday the cats were to arrive at Orly airport in Paris, they telephoned Pan Am to check for their time of arrival. They were told the "shipment," had arrived the night before and then gone on to Rome. The cats were due back in Paris at 3 p.m. that day. The airport cargo people sputtered and sweated and assured a very angry and indignant Marjorie that the cats would be fed and cared for on their unexpected leg of the journey.

They found the cats without much trouble. April, the calico, meowed when she saw Marjorie, peed in the carrier and continued to purr non-stop. Sinbad stayed in the box, nonchalantly washing himself. The group drove to the small hotel in the country, hoping to find the second batch of tranquilizer drops which Marjorie's daughter had picked up from me and mailed. They had not arrived.

Marjorie and Peggy succeeded in tracking down the veterinarian that Peggy had used in her own smuggling operation several years earlier. He dispensed tranquilizer pills and recommended a range of dosages (from one-quarter to one-and-a-quarter pills) to give the cats in trials to see how much was needed to accomplish the desired effect in each cat.

Before crossing the channel by ferry back to England,

Marjorie and Peggy planned to render the cats quiet and immobile and place them in baskets on the floor of the back seat of the car. They then would scatter all sorts of objects over them, which would allow plenty of air to get through so the cats could breathe normally. In addition, they planned to put smelly cheeses in with the cats to disguise any odors should the cats eliminate.

At the hotel, Marjorie and Peggy tried the tranquilizers on their contraband. The smallest dose did nothing except make Sinbad bad-tempered, like a nasty drunk. Doubling that dose caused Sinbad and April to sleep for a while.

Next came the strategy of the trip: which ferry to get on the following night, when to give the tranquilizer for the best effect, and how to arrange the architecture of concealment. Peggy bought four baskets, two large frying pans, a casserole dish, sketching equipment, trays, and a cardboard carton—all building blocks of camouflage.

They were told by Madame, the hotel owner, that the ride to the Boulogne ferry port would take four hours, so the two women set off in time enough to make the 8:30 p.m. ferry. En route they stopped to eat at a picnic table alongside the road, and allowed the cats out on leashes. The cats were given what they felt were optimal doses of the tranquilizers, and ate while they took effect.

Suddenly, Peggy noticed the time—they would barely be able to make the ferry! Back on the highway, Peggy sped to make up time. After five minutes, Marjorie noticed she was missing her loden-cloth coat and insisted on returning to the picnic spot. It wasn't there; she guessed that she had placed it on the roof of the car and it had blown off en route and was picked up by a passerby.

Off they sped again. Traffic was terrible. There were a number of large trucks, and their British steering wheel was in a bad place for visibility while driving on the right side of the road, making it difficult to pass other cars. Meanwhile, the cats had awakened and were drunkenly bouncing around all over the back of the car.

As they entered Boulogne, the time was 8:10 p.m. Peggy drove, and Marjorie, giving up on the pills, administered into the cats' mouths some of the drops that had arrived from my office a few days earlier. They quieted down in about 10 minutes. Peggy tried to follow signs to the ferry while driving over wet, poorly lit streets, many of them virtually pitch black.

Suddenly there was a bump and a sensation as though a knife was pulled along the underbelly of the car. Marjorie and Peggy found themselves sitting in the middle of a railroad station—on the tracks! Marjorie thought, "That's the end of the car. What will happen when we unload it?" The cats, incredibly, were asleep. Not a sign of them.

Peggy ran to the platform, and fortuitously found a representative of the British Automobile Association, who looked the situation over, left, and came back with five other men he had rounded up. They lifted the small car and turned it around in the direction of the road. The Automobile Association man then checked the auto and declared it okay.

By the time the women arrived at the port, it was 9 p.m. They had missed the ferry. Generally, the 8:30 p.m. was the last ferry of the day, but due to the Easter holiday, there was to be a special crossing after midnight.

The duo headed to the town center for dinner. When they arrived back at the car, both cats were up and active and meowing. April behaved like a drunken lady. Scouting for a dark quiet street to get some sleep, Marjorie once again administered tranquilizer drops to both cats and put them in canvas zippered bags.

When it became time to board the ferry, the cats were quiet. Marjorie and Peggy had to stop the car to have their passports checked through open windows. They wanted to keep the windows closed in case of meows (leaving one a little open for air), but upon opening them, the women became alarmed that the strong smell of the cats would give the game away. They forgot their camouflage of French cheese.

The area on the ferry to which their car was assigned was dazzlingly well lit, and passengers were told to leave the doors unlocked. "What about our jewelry?" Marjorie and Peggy asked the attendant, as though they had some. "Take it with you," was the gruff response. The two went upstairs to the lounge and were not allowed to return to the car for the hour and a half trip. Without much success, they tried their best not to think of the dire possibilities of disaster.

Five minutes before docking time they went down to the car. The worst possible thing had happened. Stretched out on the back ledge next to the rear window, in the brightest lights, lay a large black cat. He had gotten out because one side of the structure had collapsed.

Both women slid into the front seat and Marjorie reached around, hauled Sinbad off the ledge and shoved him beneath the hiding place of pots, pans, baskets, and cartons. The deck hands gave the women peculiar looks, as though they understood.

Once off the ferry, knowing the customs inspection still had to be faced, they panicked.

"Let's declare them. It's hopeless," said Peggy. They both knew the penalties for discovery: prison for themselves and euthanasia for the cats. It was obvious that they hadn't intended to declare the cats, they didn't even have health certificates. But they needed to keep moving, and found themselves at the checkpoint, agreeing to play it by ear.

They were told to get out of the car, and Peggy did all the talking, telling the inspector, a tall, young dark-haired Scot, all the things they had bought in France—an intentionally large number of undutiable items. As they held their breath, he finally said all right, made a mark on a piece of paper, and gave it to them to surrender at the exit. They were safely home at last!!

It took two hours of travel away from the coast until Peggy and Marjorie were able to laugh about the humorous things that happened. Peggy was dropped off in London and Marjorie continued up to the Cotswolds. The cats were hard asleep until they reached their new home, where they lived many more years.

Marjorie was still in a state of nervous shock and exhaustion when she wrote me about it a week later. Both she and Peggy vowed they would never, never do anything like this again, or recommend it to anyone.

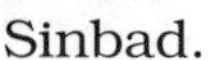

Sinbad.

April.

The Italian Visit

Following graduation from Veterinary School, I became a member of the faculty at the University of Rhode Island, College of Agriculture. Most of my work was at their Agricultural Experiment Station. The position entailed research projects involving diseases of farm animals, mainly poultry and dairy cattle; lecturing to groups of poultry farmers on various aspects of prevention and treatment of poultry diseases; and teaching a course in this same subject to senior agriculture students majoring in poultry husbandry.After two years, I took a six-week vacation to Europe, traveling alone. At one point, I found myself in the beautiful old city of Pisa. I climbed the leaning tower, visited the baptistery and explored the town. I decided to visit the Veterinary School at the University of Pisa. It was my first trip abroad, and I had an urge to meet foreign members of my profession.

On any foreign trip, I will knock on the door of a veterinary practitioner and try to chat awhile. I say try, because he or she may be too busy or we can't communicate too well because of language differences, but they always seem delighted to meet a foreign counterpart. Thomas Wolfe once said that a truck driver in America has more in common with a truck driver in France or England than with other Americans of different educational, social or economic statuses. I think this holds for professions too. There is a strong common bond between people who have undergone the same professional training and do the same work. This is especially true of veterinary medicine, since there are relatively few veterinarians in the world and it is a singular breed of person who chooses my profession.

I approached the tall iron gates of the University of Pisa Veterinary School. The buildings' colors were the lovely hue found all over Italy, showing age and softness: yellow-tan pastel-colored buildings with red tiled roofs. In the yard, a man dressed in an immaculate white coat seemed to observe a dog on a leash. I took him for a faculty member and conveyed that I was a "veterinarian Americano." He told me in a solicitous manner that he was an attendant and would take me to a faculty member.

The professor turned out to be a very personable, neatly dressed, friendly and soft-spoken young man of medium height, with brown eyes and dark hair. He was genuinely pleased to meet an American counterpart. We conversed despite the language barrier, he using his excellent French and very rudimentary English (which he was pleased to be able to practice), and I trying my scant knowledge of French. After we had exchanged some information, such as our names and the work we engaged in, Arturo offered to give me a tour of the school.

In short time, we walked through room after room of jars containing pathological specimens preserved in formaldehyde: tumors of all kinds, diseased tissues, parasites, and malformed fetuses, etc., ad infinitum. He then asked me a question, which I mistakenly interpreted as, "Do you want to see more of this?"

"No," I replied. I'd seen enough pickled specimens in my four years of veterinary school to last a lifetime.

He immediately looked crestfallen and phrased his question in another way, "Do you want to meet my professor?" I lost no time in assuring him that I would love to.

His professor, a distinguished man in his sixties, greeted me and spoke through our interpreter, Arturo. It was then that I saw the current issue of the A.V.M.A. Journal on his desk. I opened the journal and identified myself to the two men present by pointing to my name, and then to myself, as one of the authors of a paper on a poultry disease which happened to be in that issue. Instant celebrity status!

The "Professore" now took the lead of our tour of the school and ushered us to the area that housed the clinical facilities for both large and small animals. As we passed through the departments, he explained who I was to each faculty member. They were all dressed in clean, starched-white laboratory coats, and each one shook hands and greeted me warmly. I had not expected this kind of V.I.P. treatment; it felt good.

In a small animal examining room, a white-coated veterinarian examined a small dog on the table. Opposite him stood the dog's owner, a comic-looking, obese man wearing a narrow-brimmed fedora with the brim turned up, and wide, bright red suspenders holding up pants that ended about four inches above his ankles. He listened intently as the professor went through his introduction of me in Italian. Suddenly, he swung around facing me and pleaded, "Veterinario Americano, whatsa wrong witha my dog?"

We all shared a laugh tinged with some embarrassment at the client's lack of confidence in Italian veterinary medicine. I assured him that he was in capable hands.With thank-yous and farewells, Arturo and I left the school. He took me to a bar and treated me to a vermouth. What we talked about then I really don't remember, but we were both young, in a profession we loved and were excited about, and enjoying meeting a colleague from across the sea.

He invited me to breakfast with him and his wife the following morning in a suburb of Pisa. I refused because regrettably I had to travel late that afternoon to Florence. We took leave of each other warmly. In the few hours we had spent together, we both felt a friendship beginning.

On the train to Florence, I sat next to a chubby, elderly gray-haired woman dressed in a very good but threadbare heavy brownish tweed suit. As I sat down, she nodded her lined, pleasant face toward me. I said hello.

"You are American?" she asked in heavily accented English. I confirmed her suspicion. She had a meager English vocabulary, so we managed with a little French when there was an impasse as we began our conversation.

"Who are you voting for?" She referred to the impending American national election. It was September 1952 and Adlai Stevenson was running against Dwight Eisenhower.

"Stevenson," I said.

"Me too," she said excitedly, and swiftly thrust her hand out to shake mine. Though she obviously was not going to vote, she avidly followed the campaigns.

The conversation soon got around to what sort of work I did. I told her I was a veterinarian engaged in poultry disease diagnosis and research.

Her eyes lit up. She told me she had a poultry farm raising thousands of broilers, and had trouble with two common diseases, Newcastle Disease and Infectious Bronchitis. As we neared Florence, she asked if I would come to her estate outside of the city. She took a postcard out of the large satchel she used as a handbag. The sketch on the postcard showed oxen pulling a cart with containers of grapes and men working in the vineyards harvesting the grapes. Under this scene were the words CHIANTI della BARONE and a coat of arms. It seemed the primary business of her estate was the winery, and the poultry secondary. She wrote her telephone number and told me what bus to get. It was left that I should call and come if I was able. Without speaking the language, I felt insecure about facing telephones or buses, and besides, I only had one day left in Florence and there was so much I hadn't seen.

Her name on the card was Marchesa Maria Bianca Viviani della Robbia. She was a marchioness, a noblewoman and a member of the family famous for sculpture in the 14th and 15th centuries, whose marble statues and glazed terra cotta pieces grace countless churches and museums throughout Italy and the rest of the world.

Looking back, I've regretted not visiting her. I have no doubt it would have been a once-in-a-lifetime experience.

Phew!!!

A client came in with a tiny skunk that had been orphaned by a car in the country. He wanted to have it de-scented. I groaned, turning pale, as the memory of another experience with baby skunks came to mind.

Several years earlier in central New Jersey, a friend had found three baby skunks whose mother had been killed by a car. He planned to keep one as a pet and had found homes for the others. He asked me to de-scent them. De-scenting a skunk involves removing the anal sacs so that they cannot spray that familiar musky skunk material. There are two sacs that lie beneath the skin on both sides of the anus. The ducts from these sacs open into the anal sphincter muscles and if they are removed intact without breaking the sac, the surgeon avoids getting the material on himself as well as dispersing it into the atmosphere. Though this procedure (especially tough on such a tiny animal) was not my forte, I set about the task.

Sacs removed on skunk number one went fine, as did it on skunk number two. My relief was palpable and I started on the last leg of this deodorizing journey. The first sac came out fine but when I removed the second, I nicked the sac slightly with my scalpel blade. It didn't harm the animal, but the sprayed material wreaked havoc on the surroundings, aromatically speaking. The more I tried to remove the material from my skin and clothing, the more they seemed to reek.

At lunch that day in a local diner, I noticed the customers were looking at each other, sniffing. I cringed inwardly, hoping they didn't notice the smell came from me. I'm sure they figured it out, because I ended up eating alone with quite a bit of room. Did you ever have an urge to disappear? I did at that moment.

Then I had to use a bus to pick up my car which, wouldn't you know it, was in the shop for repairs. Fellow passengers sniffed, nostrils flaring, annoyed, disgusted. I tried to act casual, looking intently at the advertisements in the bus, but the olfactory detectives must have honed in on the source. I found myself, once again, with room to spare, and my fellow passengers' eyes throwing darts in my direction. I couldn't wait to retrieve my car for the ride back. How wonderful it would be to get back home, strip off my clothes, toss them outdoors, and bathe!

I promised myself never to do the procedure again. Yet there I was, once more presented with another baby skunk.

"This is not my favorite surgical procedure," I said, relating to the client my previous experience. Perhaps you can try someone else with more experience.

"I've been turned down by two other vets. Please give it a try."

He was a nice person, and I recalled that five out of the six sacs removed came out intact—an 83% success rate. So you guessed it—another story to tell.

My office was overwhelmed with skunk scent when I nicked one sac. It is singularly amazing how so small an organ holding such a minuscule amount of material can spread so rapidly and cover such an enormous area. I turned the air-conditioner fan to the exhaust position and unleashed the elixir onto the streets of Brooklyn Heights. It scented the atmosphere for several blocks around—but especially in the vicinity of my office. Most people didn't know what it was or where it came from. "City slickers," I thought.

But not Shirley Morris, a small, slender, good-natured woman who was a client, neighbor and good friend. Shirley was raised in Vermont. She later told me she had hung around my corner for a while because the smell reminded her of her childhood. I like it too, but only when it is very dilute, emanating from a distant place. Never close up.

Once again I vowed never to de-scent another skunk. And to this day, I've kept my oath.

Deadbeats and Hangups

Every veterinarian has deadbeats and hangups. Luckily they are a very small percentage of his or her total clients. People who leave their animals for medical or surgical care, collect the animal, and never pay the bill are deadbeats. Clients who not only do not pay the bill, but also abandon the animal they left with you are called hangups. One can have a deadbeat without a hangup: they've picked up the animal but won't pay. Rarely is there a hangup without the owner being a deadbeat as well—that is, one who pays for the care but abandons the pet.

* * * * *

Duke was a large, beautiful, black-and-tan German Shepherd about five years old. He had drunk a substantial amount of antifreeze and was prostrate and moribund when he was carried in on an improvised stretcher. The owner left him with us and claimed he had not brought any money or checks with him. The dog meant everything to him, so rest assured we'd be paid, he said. He'd rushed out of his house so rapidly and unexpectedly that he'd forgotten his wallet.

Antifreeze or ethylene glycol poisoning is common in dogs and cats, mostly in late fall when car owners and garages drain car radiators to prepare for the coming winter season. Animals drink antifreeze because it has a sweet taste they seem to find delicious and irresistible. Uremia and death can occur within 12 to 36 hours if the poisoning is not treated early. Therapy consists of injecting two different medications intravenously every four to five hours for ten treatments.

Mr. Gonzales, Duke's owner, was a short, stocky, muscular man in his forties with a lush growth of mustache and a deep tan, probably acquired working on the docks in his Red Hook neighborhood. He assured us that he'd be back the next day with a deposit. No, he didn't have a telephone, but he would give us his brother-in-law's number. The brother-in-law could always get in touch with him.

Next day, no Mr. Gonzales. Five, ten and fifteen days later, no Mr. Gonzales. We tried the number he had given us, but it turned out to be phony. I was sure we couldn't count on the veracity of the address.

The only positive aspect of this case was Duke's recovery. It felt good to see him alert, lively, and eating well. His blood and urine laboratory values for kidney and liver functions were normal.

Now to find the ardent dog lover who deserted Duke. I was furious at this larcenous bum. I couldn't wait to confront him, not only for cheating me out of a fee, but also for so casually abandoning a helpless creature who depended on him.

My associate, Dr. Rich Turoff and I put Duke in my car and set out for Red Hook. Many of the buildings in this area have heavily grilled storefronts on the street floor. The building fronts vary from brick to wood to asphalt and faux stone. This house, a three-story walk-up with six apartments had a hallway so dark we could only read the names on the mailbox by lighting a match. There was an overall stench of urine, with overlays of a medley of cooking odors. The paint peeled off the walls, revealing multicolored layers. Each mailbox had anywhere from two to five different names on it, none of them Gonzales. Looking for Mr. Gonzales in Red Hook is like searching for Mr. Smith in Greenwich, Connecticut.

We rang a bell and got communication from the second floor landing. It sounded like Spanish, which neither Richard nor I spoke or understood.

"No habla Espanol," I said, which exhausted my Spanish vocabulary. "I don't speak Spanish."

The speaker somehow got across to us that he thought Gonzales, he of the large dog, might live two houses down.

Again no Gonzales on the mailboxes, and again we rang a bell at random. This time, a mustachioed man came out on the first pitch-black landing. The man said he didn't know Gonzales or recognize the dog, though Duke obviously recognized him, wagging his tail and emitting joyful yelps. Getting very testy, the man left the landing when we tried to question him further.

On the street again, we cruised up and down both sides, asking everyone we saw if they recognized Duke, and hoping Duke might recognize his home. No luck. Duke seemed to enjoy the stroll as an adventure, his tail wagging, and periodically barking joyously.

The weather was cold; it was getting dark and our evening office hours were soon to begin. The quest was beginning to feel like one of those anxiety dreams—the kind where you're back in school or some other place and time, and you're in some endless, helpless, yet very urgent search for something that will be crucial to your future but you are frustrated at every turn. No one you appeal to can or will help you. And, about to panic, you wake up in a sweat and are relieved and thankful it is only a dream.

We decided to give up the hunt. The man with the mustache came outside, and spotting us, shouted in high decibels something about calling the police if we don't stop hounding him and others in the neighborhood. Rich and I concurred. There would be no use in pursuing this further. At least we gave it a good try.

A few days later, a young couple came in looking for a nice dog to adopt. Duke happily fit the bill.

* * * * *

Listen to this one. A new client, Mr. Ruiz, brought in a male Weimaraner, a large, completely grey, long-eared dog about three years old that had been run over by a truck in his own auto repair and body shop. Mr. Ruiz, a thin wiry man with a pencil-thin mustache, and grease-streaked shirt, and suspenders, described what had happened.

"Doc, it was a crazy accident. Chico was taking a snooze behind our towing truck. We got a call to pick up a car wrecked in an accident, and my brother got into the truck to back out of the garage and ran over Chico. Please try to save his leg. Chico is a valuable watchdog to us."

Chico was in bad shape. He was in shock. His left front leg appeared to be crushed. The leg was laid wide open and looked like an anatomy lesson. I could see muscles, nerves, blood vessels, bones, tendons and ligaments, as well as a generous amount of dirt, grease, metal filings and other substances found on a garage floor. Incredibly, no bones were broken. Ruiz left the dog with us along with a moderate deposit.

We brought Chico back to a more lifelike state with intensive care and then started on the limb. What a project! We had to keep debriding—that is, scraping away and snipping off tags of ruined tissue, flushing and cleaning repeatedly, and then suturing tissues and fragments of muscles together where we could, to try to get this flattened, shapeless thing to resemble a normal dog's leg.

Chico had to be anesthetized each time we treated him. In addition to the debridement and suturing, we painted his leg with scarlet oil (like that I had used on my first horse patient), a very effective liquid that stimulates new tissue growth. After several days of treatment, we sent the bandaged dog home with instructions to bring him in weekly for the treatment, as described. This went on for two months.

Ruiz occasionally paid something toward the bill but always managed to stay considerably behind.

Finally, the treatment was over. Chico's leg looked almost normal and he was all but devoid of a limp. His owner was pleased and promised to pay the balance very soon.

After six months of billing with no response, after numerous telephone calls yielding nothing but excuses, empty promises and downright lies, Rich and I once again set out on a mission; this time not to reunite dog with owner but to collect fees earned.

We arrived at his address and looked into the maws of a huge and apparently prosperous auto repair garage, though no one seemed to be on the premises at the moment. Coffee break? A passerby directed us to a busy shoe store up the street also owned by Ruiz. The woman in charge (Ruiz' wife?) told us she didn't know where he was or when he'd be back. Deadbeats often have a network of misinformation to help them in their avoidance efforts.

Rich and I walked back to the garage to take another look. Out dashed two dogs, barking, snarling, fangs bared and heading straight for us, ready to turn us into shredded dog food. One of the creatures was Chico the Weimaraner, the beneficiary of our ministrations. He showed neither lameness nor gratitude. The other was a huge Doberman Pinscher we'd seen in the office before.

Rich is a large and fearless man when it comes to dealing with humans. However, he and I were terrified and frozen with terror because of the sudden and unexpected nature of the attack. Rooted to the spot, we expected to be mauled by the two beasts. A signal was shouted from inside the garage; the dogs stopped just about a yard from us, turned, and re-entered the garage. Still, no one came out. Mouths dry and shaking, we left. Ruiz has gotten his message across: "I'm not paying the rest. Stay away from my premises."

We had a small claims court summons sent by mail. It was refused and returned to the court. We decided not to pursue it further.

Veterinarians are disgusted and angry at such behavior because it fuels our somewhat jaundiced view of a good-sized segment of mankind. Some low-life people just have to get away with something at others' expense—and they do. It is infuriating. Sometimes it really gets to us.

Despite his crude way of avoiding payment, Ruiz was satisfied with the medical results. To this day, he recommends quite a few clients to my clinic. Can you beat that? Luckily, they pay their bills.

* * * * *

Rottweilers are stocky, strong, black-and-tan dogs with large heads and powerful jaws that can crush bones without half trying. The owner of one particular Rottweiler appeared in my office, glassy-eyed, shabby and seemed in a chemically-induced other world. Mr. Jones came from an area known for a high crime rate due largely to drugs, and I suspected Jones was under the influence.

"Doc, this is my pal Max. He's an attack-trained dog. Someone tried to rob me last night and I set Max against him to protect me. Max bit the thief, but he got stabbed a few times." Attack-trained meant that on signal, with a special word or phrase, this dog could turn from placid pet to killer. I hoped that while he was in our care, we wouldn't accidentally utter the signal.

On checking Max, he did have several deep stab wounds in the brisket area (the well-muscled front part of the chest) and a couple on the neck. Mr. Jones, of course, didn't have so much as a coin with him. Since surgery was not going to be done until several hours later, he was told to return with at least a deposit.

"Doc," he implored, "do you think I would not come back for him? He saved my life. He means more to me than anyone. I'd never walk out on him." Where have we heard this before?

We never saw him again. He left no telephone, a phony address, and probably used a fake name. It was Jones' loss and woe if he would have another encounter with a thief. Max turned out to be a real pussycat of a dog. A lover. We fixed him up and three weeks later found him a good home. I hope his new owner never stumbles upon the code word, or that Max forgets it because of disuse.

* * * * *

The next one happened to a colleague many years ago but I am including it because of its humor.

Dr. Milt Firestone had a small animal practice in the Bronx, the northernmost borough or county of New York City. Hospitalized dogs were let out to exercise one at a time, in a well-fenced area behind the clinic. On one occasion, he had a male short-haired terrier, mostly white with a few black patches, including a black patch over his left eye. He looked like the terrier used in the Victrola advertisement many years ago—"His Master's Voice"—remember?

Anyhow, Milt had Zippo for two weeks, treating him for a very serious illness called distemper. He had not seen a dime in compensation yet, though the owner kept calling to keep up with Zippo's progress.

Zippo was finally back to his old lively, bouncy self and ready to go home in a day or two. Milt let Zippo out in the exercise area and returned to the work he was doing inside the office. About half an hour later an assistant went to bring Zippo in. The yard was empty. Milt's stomach experienced that horrible queasy feeling one gets when looking down the parapet of a tall building and thinking of the impending doom.

He and his assistant set out to scour the neighborhood. "How could this little dog have gotten over such a tall fence?" he thought. It was almost impossible to do without help.

After two hours of searching, Milt rounded a corner, and, sitting on a bench in front of a wooden-frame house was a tall, thin man: Zippo's owner. And sitting next to him was Zippo! No other dog could be marked exactly like that.

"Mr. White, thank goodness Zippo found his way back to you," Milt exhaled. "I've been going crazy looking for him since he disappeared from my yard."

"This ain't Zippo," said White. "This is his twin brother." Milt never did get paid. Obviously Zippo had help in escaping.

Rupert

I had never been to the casinos in Atlantic City, or any other gambling establishment, and was curious to see what they were like. My friend Charlie made arrangements for us to go on a Wednesday, the day the clinic was closed for thorough cleaning.

Charlie and I went by bus, and deposited modest sums at a couple of casinos. We got a taste of the glittery atmosphere that exists only in casinos; garish lighting, hordes of people in a blue-gray acrid haze of cigarette smoke putting money into slot machines with the resultant pinging, ringing and whirring noises they produce. We returned to Manhattan broke and hungry about 9 p.m. and decided to get some food.

When I telephoned my wife to tell her where I was, she sounded upset and said that something had happened and that she would tell me about it when I got home. So much for eating. With foreboding, Charlie and I caught the subway and headed back to Brooklyn Heights.

My office had been robbed!

While I was trying my luck with fate in New Jersey fate was playing with me in Brooklyn. The loss at the casinos was negligible; the loss at the office was considerable—financially, by inconvenience, and by instilling fear and anxiety in victims directly involved. Fate played a part throughout the saga . . .

The events took place as follows: George Powell was

the person who thoroughly cleaned the office on Wednesdays. He was a gentle, tall, thin, black man in his mid-fifties with many missing teeth which distorted his southern Virginia drawl. He did his job as usual that day.

Generally, there are a few people who come by when we are closed. They either haven't checked our office hours, or attempt to get some information or medications. For those who need emergency medical attention for an animal, we provide information directing them to another animal hospital on our telephone answering machine. George was instructed many times to ignore knocks on the door, since no one around can help with veterinary emergencies. But George had left the door partially ajar and three young men pushed it open. They walked in, demanding to know where the money was.

The first, in his mid-twenties, was thin, swarthy, dark-haired, and of medium height. The second, about twenty years old, was taller. He too was lean, but had lighter hair. Both were dressed like policemen! Each wore a light blue shirt with epaulets and brass insignia on the collar, navy blue ties and trousers, and black shoes. The third man was short, stocky and barrel-chested. He was dressed in a worn T-shirt and denim jeans.

George didn't know anything about money. "I just clean here," he said,trembling with fear. There wasn't any money in the office, in any event. Angered, the thieves handcuffed George to a kennel door and stole forty dollars from his pocket (which he could ill afford to lose). They cut the office telephone wires and proceeded to ransack the office, looking for things to steal. They took jars of antibiotics, which may have been mistaken for narcotics; they took our fresh supply of postage stamps; and, worst of all, they stole all of our surgical instruments—even our duplicate sets. They took everything.

Meanwhile, in my apartment above the office, Bernice noticed that the extension of the office telephone had not rung for a while. Even though messages were picked up by the clinic's answering machine when there were no office hours, it still rang upstairs—and it rang very often. Bernice went down to check on this, and seeing what appeared to be policemen, said to one of them, "Officer, can I help you?"

Fate played a hand in this too. Bernice had never gone down to the office to check on the telephone before.

The "officer," an off-duty security guard, pulled a knife and held it to her throat while one of his cronies tied her hands behind her back using telephone wire. She was blindfolded with some gauze bandage, but only after she had a good look at them. They removed her wedding band and diamond ring, and a chain with a gold charm from around her neck. She was terrified, as was George, still cuffed to the kennel. Then they stole the item that would do them in—a Blue-Fronted Amazon parrot named Rupert.

Rupert used to live in a large cage with an open door in my apartment, but I moved him downstairs after he destroyed a number of coveted books and bookshelves. His new home was a large kennel, complete with a nice log for chewing, and lots of animal and human company.

The short stocky thief decided to steal Rupert. When the three hoodlums were finished with the robbery, the stocky thief walked out with the bird on his shoulder. Rupert didn't know the difference between the good guys and the bad guys, and he didn't offer the slightest resistance. How many thieves out for money and valuable loot would steal a bird?

After the trio had left, Bernice extricated herself from the bindings around her wrists and immediately called the police, giving them descriptions of the three men and the parrot.

Information went out over the police radio, and

serendipitously, as the message came through, a foot patrolman walking along Court Street about one-half mile from the office saw a black, dilapidated Volkswagen beetle go by with three men in it, one with a parrot on his shoulder.

The officer hailed a passing patrol car and gave chase. When they caught up with the Volkswagen, the uniformed duo, the ones Bernice and George mistook for policemen were in it. The duo was picked up in front of a pawn shop on Atlantic Avenue about to dispose of the jewelry. The thug with the parrot and all the heavy loot had "jumped ship" somewhere along the route.

At the precinct, the two men claimed they knew nothing about a robbery, a parrot, or a third man. A thorough examination of the back seat of the patrol car revealed the wedding band and engagement ring, which had been secreted behind the seat cushion. The chain with the gold charm, a strawberry that a talented jewelry-maker friend had fashioned, was never found, but my wife was happy to have at least the rings back. However, since they were evidence, they were kept at the police precinct for quite a while. Although we hated to leave them, there was not much we could do.

That afternoon, two detectives from the Precinct were at my office taking notes, questioning Bernice and George about the events and taking down descriptions of the thieves. They were still around at 9:30 p.m. when I arrived home, and I was better able to assess what was missing. The instrument loss hurt. Like tools in any field—art, auto mechanics, carpentry, cabinet making, etc., the practitioner accumulates tools that just work perfectly for him, built and designed just right and often hard to duplicate. I somehow hoped to get them back, but realistically sensed a loss.

Later that night, at the precinct for more questioning, we overheard a young woman relating to an officer at a nearby desk that a man she knew sold her a parrot that day, and knowing the person, she suspected it was stolen. Apparently an attack of conscience had overcome her concerning the owner of the missing bird. Another coincidence, or was it fate once again?

I jumped up and approached her. We recognized each other from the clinic. Her dog was a patient of mine. After I told her what had happened and described the parrot to her, I went back to her apartment and retrieved Rupert, no worse for the experience, and neither joyous nor excited at seeing me. So much for parrot loyalty.

My client did not know where the suspect was, nor did she know his name. She only knew that he was a neighborhood hoodlum, involved in scrapes with the law for many years. Purse-snatching, drug crimes, and burglaries were his line of work. He was aggressive and working on a career in criminality. She knew nothing about the surgical instruments.

A Brooklyn colleague and friend, Dr. Syd Kessler, loaned me some surgical instruments, which enabled me to carry on temporarily.

The two apprehended thieves, brothers in fact, were out on bail almost immediately and went back to their jobs as security guards at Madison Square Garden—talk about foxes guarding the hen house!

In the ensuing weeks, several aunts, cousins and family friends of the sibling criminals tried to strike a deal when they came in with their animals: would we drop charges against those two upstanding citizens if they could somehow retrieve and deliver the stolen property? The answer was no.

Finally, the identity of the third member of the group was known. According to police annals, he had a record of crimes, arrests and releases which ran to five pages in length. Drug use and sales, muggings, burglaries, robberies—a dedicated career criminal in his early twenties. He was nowhere to be found.

Bernice went to the precinct and identified the brothers in a lineup. She persisted through various hearings and numerous postponements, all initiated by their lawyers in an effort to discourage her from going on with the case. The brothers finally plea bargained. We don't know if they ever spent time in jail.

The third thief was caught in Florida, where, by that time, he had perpetrated numerous other crimes. We believe he served time.

Meanwhile, during the many weeks these proceedings went on, the two rings remained at the precinct. Having heard that things have a way of disappearing and the chances of disappearance are greater the longer they stay, we prevailed upon a friend in the district attorney's office to see if he could get the rings released. Lucky for us, he did.

The instruments were never recovered—probably sold for a pittance. I had never thought to insure them, and the cost for replacements was over $4,000. Who ever imagined someone would steal such material? Had the thieves not taken Rupert, I doubt any of them would have been caught.

George was reimbursed for his loss. I was relieved and thankful that Bernice and George hadn't been harmed.

One doesn't need Atlantic city for games and losses. One can have them at home. Every day is a gamble.

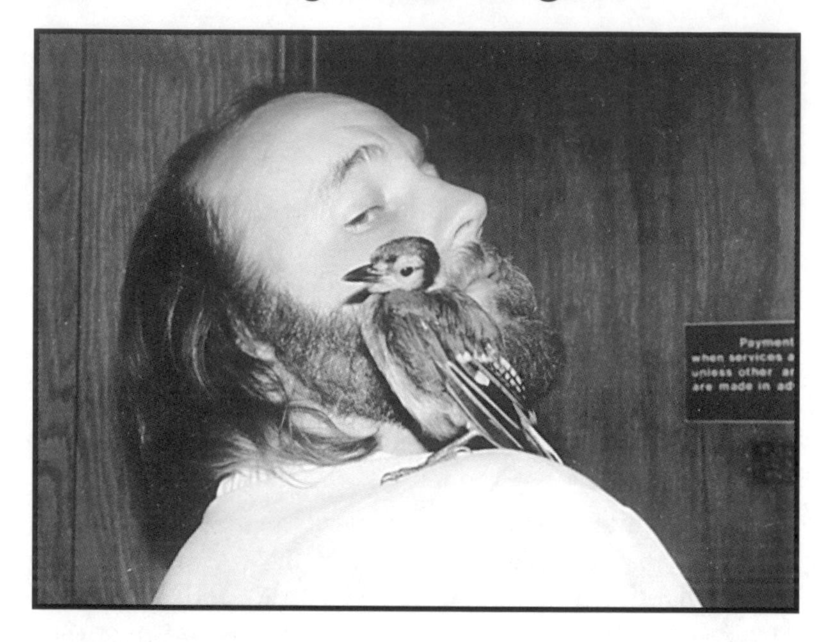

Above: Dr. Rich Turoff with orphaned bluejay.
Below: With pigeon.

Above: Rupert.
Right: Bill Hinkle with Rupert.
Below: Rupert in the getaway car, as seen in *The Brooklyn Paper.*

The Brooklyn Paper / Batton Lash

Other Creatures

Years ago, pet shops sold newly hatched ducklings at Easter time. Buyers for such adorable little birds were plentiful. Many pet shop owners couldn't care less about informing their customers about duck husbandry, and therefore the customers took them home, and hadn't the slightest notion of what to feed them or how to care for them.

These ducklings were primarily fed bread, plus odd table scraps. Within a couple of weeks they were scrawny and "down on their hocks" because of tendon problems due to magnesium and other mineral deficiencies. Their hock joints, similar to our ankles, were unable to extend. They walked with great difficulty. Additionally, once their owners tired of cleaning up the cute little creatures' bowel movements, veterinarians could count on a number of ducklings brought in for euthanasia a few weeks after Easter.

I was unable to kill these birds, so we kept proper poultry food on hand and accommodated the ducklings in a large kennel where they were well fed and watered.

At one time, our country home was used as a pheasant farm. Pheasant eggs were hatched in incubators. The chicks were raised under brooders to keep them warm and then put in a pen in the barn. The pen had a small swinging door, not unlike the type used to allow small dogs and cats to go in and out of houses. This door allowed the growing baby pheasants to go outdoors into an area penned with chicken wire to get air, sunshine and grass. The barn was unused for several years before we obtained it.

That spring and summer Bernice stayed on the farm with our two preschool sons. I drove up midweek for the day, and Saturday after office hours for the weekend, and brought up any ducklings that had come in during the week. After about a week on good nutrition, they walked better, and acted stronger and livelier.

When chicks and ducklings are hatched in an incubator as opposed to hatching under a mother duck or chicken, they attach to and follow the first moving creatures they contact, and these creatures become their mother. Mother could be a human being, a pussycat or whatever. It is called imprinting.

In this instance, they chose my four-year-old Harvey. Both the ducklings and the bantam chickens we kept were given the run of the property during the daytime, and the ducklings wouldn't leave Harvey alone. Wherever he went, four or five ducklings waddled after him in a straight line, quacking away.

Though it was a source of great amusement to observers, somehow it caused enormous frustration in Harv. He couldn't make a move without them during the daytime. At night the chickens automatically went back into the pen, but the ducklings were carried in. I was pleased to see how sleek and healthy they were growing. Adversely, Harvey (my little duckling) would sometimes cry because of this avian impediment to his activity. I tried to show him the cute and humorous side of it, but he clearly felt his freedom was being encroached upon.

Unfortunately, this interlude did not last too long. The small swinging door to the pen, unused to this point for many years, was accessible from the outside. One morning, upon opening the door to the pen to let the birds roam and munch on grass and insects, I discovered that all the ducklings had been killed! An animal, possibly a ferret, had entered through the small swinging door and eaten the poor ducklings. All the bantams were spared since they were able to fly onto roosts and nesting boxes and out of harm's way.

The sight of the poor victims saddened and nauseated me, and engendered a terrible feeling of guilt for having neglected to think of possible wild animal entry through that portal. Although Harvey had expressed negative feelings about parenting these ducklings, he too was upset. He seemed to miss them, perplexed by the mystery of death, especially sudden death, which he had not encountered before.

Of course, I nailed the "barn door" closed after the damage was done. The duckling season was over. I believe there is now a city law prohibiting pet shops from selling ducklings and chicks.

* * * * *

Louie was a baby squirrel that someone found at the base of a tree and brought into the office. His mother was nowhere to be seen, even after a thorough search and sustained observation of the site. So our technician made a mixture of ground nuts, honey and milk and fed Louie at regular intervals with a syringe. We intended, when he was fully grown, to release him in some nice green park where there were many other squirrels. He lived in one of our kennels and was a sweet diversion for all of us when we had a little free time.

Though he was little enough to be tucked into a breast pocket, sticking out his head and showing his large brown eyes, Louie preferred to race around than to remain still. He was almost perpetually in motion.

On one occasion I took him up to my apartment. He made a nest on one of my bookshelves and secreted nuts there. He also gnawed my shelf, as well as several books. He literally flew around the apartment at times, driving my two cats crazy. I had to capture him in midair using a net to return him downstairs. It was almost like catching butterflies, except that this baby squirrel was a supersonic jet.

Sindy Habib, our receptionist/technician, loved playing with him. Noticing that he was rather quiet one day, not behaving like a hyperactive rodent, she picked Louie up and was bitten on her hand. Never had Louie done this before.

Shortly after that, Louie died. Though it was extremely unlikely, we had to consider rabies, even though he couldn't have been bitten by a rabid animal, because we had gotten him fresh out of a nest. In addition, there hadn't been a case of rabies in the city for many, many years up to that time and I had never heard of a case of rabies in a squirrel anywhere. Squirrels move too fast to be bitten, and as vegetarians, they wouldn't get it by eating infected animals, as it may be transmitted in the wild.

Nevertheless, taking no chances, the New York City Health Department Laboratory checked poor Louie's brain for the rabies virus. It was negative.

We concluded that Louie had probably picked up some insecticide we usually put out above the kennels when Sindy had let him run free in the kennel room. He was a cute, fetching little guy, and we missed him.

* * * * *

Another client brought a patient in to be looked at, a sweet, healthy, young black-and-white Belgian rabbit named Daisy with a crushed toe. We told the owner that amputation of the toe was necessary and that the bunny would be as good as before. Though she was quoted a reasonable fee, the client decided to put the money to better use and have Daisy put to sleep. I believe she wanted an excuse to get the animal out of her home: she was obviously able to afford the fee, and probably bought Daisy in the first place to please a child.

We performed the surgery and Daisy recovered without ill effects. We decided to keep her.

All of the employees in the clinic enjoyed having the sweet presence of Daisy in the hospital, and in between office hours we let her have the run of the place. One day a large black male rabbit was brought in with a fractured bone in his right forelimb. We anesthetized, radiographed him, and set the forelimb in a cast. Once he was wide awake, we put him in the kennel with Daisy for a couple of hours, with his owner's permission. He went home the following day.

About four weeks after this conjugal visit, Daisy was growing more and more bald. Watching her, we noticed that she was biting her own fur out and making a nest. She was expecting!

Rabbits gestate between 29 and 35 days. A few days or hours before kindling (the term for giving birth in rabbits), they remove fur from their forequarters, breast region and hips to make a nest.

A couple of days later, Daisy gave birth to six bunnies, pink and bald. Shortly thereafter, they were covered with black-and-white silky fur. We supplied bunnies to several kindergarten classes of schools in the neighborhood, and a rabbit fancier we knew took Daisy.

Not bad for a rabbit who supposedly was dead for several months.

Benjie

I have lived with quite a number of cats over the years, but memories of my first cat are strongest in my mind. You might think it is natural to remember a first cat. But it isn't just because the cat is your first. My cat and I lived alone together; he was my source of joy and companionship.

Early in my small animal medicine career, I worked in an area of Brooklyn called Flatbush, at an animal hospital owned by a colleague, Dr. Syd Kessler.

One day during clinic hours, a woman brought in an eight-month-old Siamese cat with a large abscess on one of his front legs. His owner was upset, and so was Tippy, as he was called then. Mrs. Mitchell brandished her bandaged hand and explained that Tippy had escaped from her house and gotten into a cat fight. She tried to break up the fight and was bitten—by Tippy!

An abscess is a pocket of pus under the skin and in the tissues beneath. It is very painful, and almost always the result of a bite from another cat.

She left him for treatment, and said on departing that if her children agreed, she'd like to give him away. Apparently, she was no great animal lover.

Tippy was seal-point in color: light tan body with dark brown ears, muzzle, paws and tail. And he had the most beautiful pale blue eyes that I had ever seen. His appearance, combined with a benign, trusting, sweet expression, hooked me.

I performed the necessary surgery to treat the abscess. In the ensuing couple of days while Tippy was in the hospital, I hoped to hear positive news from his owner. Not only was he an appealing animal, but I was ready for a pet. I had switched jobs several times from other branches of veterinary medicine (research, teaching and diagnostic work on farm animals) and so had moved around a bit. I was now living alone in a small apartment in Manhattan.

At last the call came from Mrs. Mitchell, Tippy's owner. He was mine! I hadn't had a pet since I was a young boy—and never a cat.

Benjie, as I renamed him (subconsciously after myself, it appears, since I discovered not too long before this event that the name on my birth certificate was really Benjamin) was a delight from the start. I looked forward to returning home after work. As I approached my door, I could hear his throaty cry, a penetrating sound that only Siamese cats can bring forth. Before opening the door, I'd call "Hi Benj," at which his vocalizations increased in volume and frequency.

After I entered the apartment, he leapt at me and ricocheted off my hip, landed on the floor and repeated the acrobatics two or three times more. I had never seen a cat do that. I loved it. He was so happy to see me, and the feeling was mutual. I fed him, and then, for some time, he'd cling to me, sitting on my lap or lying on my chest face to face, purring with contentment all the while. It was soothing.

Benjie took time out from grooming himself with his rough tongue to do the same for me, on my face (ouch) and my hair. Although this form of attention wasn't exactly welcomed, I felt that Benj thought he was doing something nice for me, so how could I stop him? I enjoyed learning all his feline ways.

Benjie loved to have mock fights with me. I tossed him into the air, as high as the ceiling, and caught him. He purred all the while and asked for more. I took him along when visiting my parents, and my father, an unabashed animal lover, enjoyed sparring with Benj, who in return would spring and attack, tiger-like, and deliver swipes with paws, and gentle, painless bites.

I transported him in a by cat carrier in my car. The first time I failed to use a carrier turned into a near disaster. My frightened Benjie was all over the car at once: underfoot, on my head, draped around my neck, hanging from the steering wheel by his front paws. I dared not try that again for a long time.

Shortly after Benjie and I found each other, I started my own veterinary practice. I also got married. My wife loved and enjoyed Benjie as much as I did. Whenever we left him in someone else's care we missed his gentle, affectionate presence. I know he missed us, because upon our return, he'd greet us profusely with his distinctive Siamese sounds, ricocheting off our hips several times and clinging to us, purring loudly.

In time he had two more companions, infant boys about a year and a half apart. Harvey and Andrew were raised very close to Benjie: in the crib, in the playpen, and in the carriage. Benjie was gentle and relaxed with the boys and we had no fear that he would harm them. I even think he would have attacked anyone who might have tried. Both sons, adults now, have cats of their own. Presumably Benjie's influence.

When he was eight years old, Benjie became very listless and weak, totally losing his appetite. Force-feeding him resulted in vomiting. Laboratory blood work indicated it was probably an unknown (at that time) viral infection. I treated him as best I could with intravenous and subcutaneous fluids, vitamin injections and antibiotics. In spite of all this, I watched my beloved friend worsen. He was dying. I panicked.

They say a physician shouldn't treat his own family. Perhaps a veterinarian shouldn't minister to his pet. Was I too close to him so that I was not thinking clearly, I thought? Was I missing something that could be done?

I took him to a colleague for whom I had a great regard, Dr. Henry Sussman, who agreed with my methods after I outlined what I had been doing. He agreed with it all.

"I'd do one more thing if you can arrange it," he said.

"What's that?"

"Give him a blood transfusion," Henry offered. "Preferably from a street cat, and preferably not young. A stray has probably contracted many infections during his lifetime, and since he is still alive, has obviously conquered them. His blood will have antibodies to fight numerous diseases."

We generally had one or more stray cats in our kennels. These were picked up in the street, subway, or any other place that abandoned cats could be found, by my animal-loving technician, Ronnie, or by me. Some were abandoned on our doorstep in a box, with an anonymous note. We got these cats in good shape and subsequently vaccinated them, altered them and found them homes. This time, ironically, we had no resident stray.

As soon as I got back to the office with Benjie, I telephoned my friend Syd Kessler. Providentially, he had a stray cat.

His stray was an older smoky-gray male, very good-natured. His many missing teeth and notched ears were signs of numerous street battles, won and lost. He was a prize winner to me.

We drew blood from Smokey, as we named him, and administered it to Benj, who in all respects was moribund. For the previous two nights, he remained in his box on our bed. Too weak to get up, he urinated where he lay. Benjie was in extremis.

During the night following the transfusion, he stood up, and, albeit wobbly, he tried to walk. He cried out weakly, letting us know he had to go to the litter box. In the morning, he ate a small amount of food, and from that time on, slowly, a full recovery took place. We took pleasure in every gain he made along the way.

As for Smokey, we altered him, vaccinated him, cleaned his remaining teeth, fattened him up and smothered him with affection. Then we found him a good and loving home—a wonderful outcome for both participants in this drama.

On the way up to our house in the country, Benjie dozed on the shelf behind the rear seat of the car. At last, he had learned to relax while free in the car. Somehow he always knew when we turned into the dirt road leading to the house; he'd get up, stretch and start talking. It never failed. We let him outdoors on the lawn, attached to a long piece of twine fastened to a tree. In this way, he could explore, nibble on grass (and throw up), and find sun or shade as it suited him. A couple of times he slipped out of his collar and wandered away, causing panic while we called out his name in high decibels. Yet, he was as happy to be found as we were to find him.

Benjie had discovered the beautiful deer mice that shared the house with us. Periodically, he would catch one, carry it gently in his mouth, release it, catch it again, and so on. Each time, we removed the mouse from his mouth, lectured him on reverence for life and peaceful coexistence, and released the mouse outside the house. After numerous such incidents, we knew we had succeeded in converting him to our point of view. It was proven one day, as we were all sitting in the living room, when a mouse scampered across the floor. All eyes were focused on our cat. He sat there, his head swiveling slowly, following the path of the mouse. He didn't go after it!

Which lesson had sunk in? Kindness and coexistence or, "what's the use, they'll only take it from me anyhow"? Nonetheless, his mouse-catching days were over.

When Benjie was 17-and-a-half years of age, we went on a three-week trip abroad. I left Benjie in the care of my technician, Ronnie, and the veterinarian who was filling in for me. On our return, I found Ronnie still at the office. It was midnight, so I knew something was wrong. Ronnie's usually smiling face was long and grim. He told me that Benj had taken ill shortly after we had left: loss of appetite, excessive thirst and urination, followed by dehydration. By now he had lost a great deal of weight. My family and I were heartbroken. The results of Benjie's blood tests indicated that his kidneys were failing. All supportive treatment needed was given: hand feeding, intravenous fluids, vitamins and antibiotics. Too weak to raise his head, he seemed to have hung on until we returned.

We took him to the country with us, and for the last time, he managed to give his small cry of recognition as we turned up our road. His only treatment now was nourishment and water amid countless hugs and kisses and copious tears. He died in my lap the next day. Tearfully, we buried him behind the barn. He had for company two miniature poodles (my sister's), a guinea pig and an iguana (my son Andy's), and a bluepoint Siamese cat named Homer (ours). For some time afterwards, we imagined we saw him out of the corners of our eyes in familiar places in the house.

Although it has been many years since his death, we think of him with pleasure and love, and we occasionally slip and call other cats Benjie. We won't forget him.

Above and below: Benjie.

Far From Routine

Pierre was a small white miniature poodle who had been brought in because his owner had noticed that his urine was red. We gently expressed the dog's bladder and saw that it was indeed quite red. Red normally means blood. However, the dipstick test showed no blood present in the urine, and all other readings were normal as well. No blood?? It was bright red! It had to be blood! Were the test strips outdated and no longer working? No, they were fresh and dry and well within the expiration date.

I examined the patient. He was healthy, lively, and, according to his owner, had shown no signs of illness. I was puzzled.

About this time in the proceedings, Dr. P.J. Field, always called "P.J.," a bright young vet who was filling in for my vacationing associate, finished with the patient he was seeing in the adjacent examining room. He had heard some of the conversation and had gotten the essence of what was going on. P.J. motioned for me to step out of the examining room for a moment.

"Ask her if the dog ate a lot of beets last night," he said.

It seemed strange to me, but I asked.

"Yes, as a matter of fact. I had prepared a large bowl of beets for dinner. It was on a counter, and before I realized it, Pierre had gotten up on a chair and eaten most of the beets. I never dreamed dogs liked beets."

That explained it. I'd never encountered this before, and I asked P.J. how he knew about this.

"It happened to me once, after I binged on beets," he said.

* * * * *

One evening during office hours, a young man hurried into the waiting room and excitedly told the receptionist about a cat lying in the middle of the street two blocks away. He seemed to think the pet belonged to someone who lived in a newly built 30-story apartment building adjacent to where the cat lay. People were afraid to approach the cat, and with good reason. Even the mildest animal, when bewildered and in great pain, may very well bite its owner, let alone a stranger. Injured animals must be approached with great caution.

I apologized to the clients in the waiting room and told them I'd be back very soon. Armed with gloves and a large cat carrier, I followed the messenger to the site.

A large black-and-white cat was lying on his side in the middle of Middagh Street, which someone had already blocked to traffic. Cars were parked alongside the curb and a score of people gathered around.

The cat hissed and growled but appeared, on cursory examination, intact. There was no sign of blood and he had no difficulty breathing. I laid the opened cat carrier over the cat, scooped him into it and closed the lid.

Back at the office, we examined him after office hours.

His color was good, which ruled out internal bleeding; respiration was normal, so he probably had not sustained a tear in the lungs or a diaphragmatic hernia. His pupils were even and reacted normally to light, so there was little probability of a concussion. He hadn't even fractured any bones, though he was slightly lame in his left front limb.

The next evening, a young couple came into the office. They lived in the building and were missing their cat, Charlie. We brought the patient out, and in seconds it was obvious Charlie was most happy to see his owners.

"How did he get out?" I asked.

"He went out on our balcony, which he always does, so we paid no attention to it. We went to bed, and when we got up this morning, he was nowhere to be found. We didn't see the notice on our lobby bulletin board until we got back from work."

"What floor do you live on?"

"The 25th."

He went down 25 stories!!! Why wasn't Charlie destroyed as he should have been from that height?

We could only postulate that since one of the cars parked alongside the building in a direct line with the cat's position was a convertible. Charlie must have bounced off the top, which acted as a trampoline. I'm sure that's a world record for a fall where the cat survived with practically no injuries.

* * * * *

With each rabies vaccination, a tag with an accession number and year of vaccination is attached to the dog's collar. The tag also indicates that the vaccination was given by the Heights Veterinary Hospital and lists the telephone number. In addition to proof of vaccination, the number on the tag sometimes serves a useful purpose. We have reunited many a lost or runaway dog with its owner. A person finding a dog with a rabies tag bearing our telephone number, tag accession number and year issued, telephones us. We look up the number in the rabies logbook and get the owner's name. The dog's chart is pulled and the address and telephone number are given to the caller after we verify the description of the dog. (Occasionally someone loses or throws away a collar with a tag on it and it ends up on a totally unrelated animal.)

We had given a rabies vaccination to a client's dog. He was a Gordon setter, a beautiful black-and-tan dog with about the same configuration as an Irish setter. One of his owners worked for a large investment company and would soon be transferred to a Paris office. The family and dog would be there for at least three years. We gave them a rabies vaccination tag and certificate as well as a health certificate to get the dog into France. Our receptionist did not record the tag number in the rabies logbook since the dog would be out of the country for several years. As a matter of policy, we do not enter dogs moving to far away places since we would not expect to be called from out of state or abroad about a lost dog, and wouldn't be able to provide the finder with an address and phone number even if we were called.

Well, it happened.

A few weeks after the Gordon setter's visit, we received the following letter from the newspaper, *PARIS JOUR*. With the use of a French-English dictionary, the following emerged.

> *PARIS JOUR*
> Paris, 12 June 1970
>
> Doctor,
>
> The Brigade for the Defense of Animals, an animal protection group which depends on the newspaper *PARIS JOUR* has collected a beautiful black-and-tan Gordon setter, purebred, about six years old. This animal is in one of our kennels where we accommodate lost dogs.

Permit us to alert you because the dog carried on his collar a rabies tag # 786 issued 1970, with your name and telephone number.

The American Embassy imparted your address, which permitted us to contact you. We are very desirous of finding the masters of this beautiful dog, who is very sad in spite of the kindness evidenced by the caretakers of the refuge.

We want to thank you in advance for responding to us by return mail, if you have the information that will permit us to find the owners of our ward. It is probable that they are searching for the dog nearby.

Yvonne Stephane

What to do? We couldn't remember the name of the client and the tag number was unrecorded. This was one time we desperately needed to connect a name with this tag number, even though we would not be able to provide the owner's address in Paris for the newspaper. We could at least give them a name to search for. Everyone at the office scoured their memories to try to remember a name. No luck.

Suddenly, I remembered that the owners were good friends of Diana Foster who sold antiques across the street from my office. I ran over to her shop.

"Diana, do you happen to have an address for your friends who recently moved to Paris with their Gordon Setter?" I explained the reason for my question.

"Sure I have it," and she wrote their names, Temple and Brigitte Von Stackelberg, their home address, and his at the investment company.

Did I dare hope? "You wouldn't happen to have their telephone number too?"

"Yes I do. In fact, I was planning to call them today. Let's see, it would be about 8 p.m. in France right now."

She called, gave them the telephone number of *PARIS JOUR* and the name of the letter-writer; they contacted the paper and retrieved their dog.

After that, we recorded ALL rabies tag numbers and asked migrating clients if they had an address at their new destination.

* * * * *

Here's another case that might have qualified for the *Guinness Book of World Records.*

A first-time client came in with her eight-year-old Beagle, a gentle tri-colored female with dark brown eyes and a tail that wagged with every petting stroke and soft word spoken to her. The complaint was that Sparky constantly strained to urinate—and constantly meant almost 100% of the time—for about two years. Her efforts produced only minuscule amounts of bloody, foul-smelling urine or no urine at all. Talk about promptness in responding to a medical problem! Holding back my anger towards this elderly dog owner, I examined Sparky.

It took only a moment of palpation in the area of the abdomen just anterior to the rear legs to reveal the cause of the problem. Her urinary bladder was so full of urinary calculi (stones) that there was no room for urine to accumulate and be stored. The lining of the organ was undoubtedly swollen, abraded and infected, hence the constant straining and bad odor. This dog had been in agony for a long time and the owner had ignored it.

I felt angry and baffled. How could a pet owner, day after day for years on end, see these obviously abnormal and painful symptoms and think nothing was wrong? Her owner was informed that surgery was necessary and arrangements were made.

A cystotomy was performed (cutting open the bladder) and literally thousands of stones were in the bladder ranging in size from grains of sand to over two inches in diameter!! The lining of the bladder was an infected, bloody, thickened mess due to the constant abrading by the stones. This had to be a record for the most stones ever removed from a urinary bladder. Sparky made an excellent recovery, and by the time the sutures were removed, she was a much happier dog. She was put on medications and a special diet to try to prevent a recurrence.

Calculi from urinary bladder.

Goats Plus

My practice is in the largest city in the United States. One does not associate the treatment of farm animals, wild animals and wild birds with a city practice but strange things do happen.

* * * * *

One Saturday morning towards the end of office hours, we received a telephone call from Joe Scotto, a client who brought his dog in regularly. Yet this time, the patient was to be a four-footed creature of a different sort: a sick goat. We told him to come in when we'd be clear of other animals and people. I looked forward to this unusual patient.

"Doc," he said, after carrying the goat in, "she had two kids a few days ago up at our farm. I didn't feel I could trust the inexperienced farmhand to take care of her if something happened so I brought her and the kids back to Brooklyn. They've been in my basement. Last night Muncher lay down, stopped eating and nursing the kids. She's been shaking and having muscle spasms."

With this history and an examination, it was obvious that Muncher had eclampsia, milk fever as it is commonly called. This is a condition of female animals, primarily dairy cows and goats (though I have treated quite a few dogs for it) that occurs after birthing. Calcium is removed from the blood stream and tissues in order to produce milk for the young. This may result in an abnormally low blood calcium concentration that leads to a variety of symptoms. The lowered calcium causes severe trembling and rigidity of the muscles of the limbs which drives up the temperature from a normal of 101.5 degrees F. to 105 to 107, hence the name milk fever. It can adversely affect the heart as well—unto death.

Slowly, we injected a large volume of calcium gluconate intravenously until the trembling and rigidity of the limbs stopped. Then, deposits of the same solution were injected under the skin to be called upon later by the body if needed. With that, Muncher was picked up from the table, placed on the floor and miraculously appeared as good as new. Joe Scotto put her rope halter around her neck and started out of the side exit. My assistant and I followed him out and observed a scene that topped off the satisfaction we'd just had in helping this animal.

Across the street, on the corner of Marcolini's Wine and Liquor Store, was the usual Saturday gathering of older neighborhood cronies catching up on the news of the week. These were retirees—bald, graying, or both—running to obesity and wearing loose, suspendered pants. One of them happened to turn around as Muncher was walking up the street and did a double take. He turned back to his group and then his head rapidly swiveled around again as though he couldn't believe he'd seen a goat being led on a leash on Cranberry Street in Brooklyn Heights. They looked on, incredulous, mouths agape, giving us all a good laugh.

* * * * *

Our second goat patient was brought in by Tony Rizzo who owned a chicken market in Brooklyn and had been bringing his German shepherd in to see me. When he put his goat on the floor, she couldn't stand. Her carpal joints (wrists in humans) and stifles (knees) were in contact with the floor. They were swollen, calloused and scabbed over—an unsightly and doubtlessly painful mess. And no wonder. Her hooves were so grossly overgrown they became gnarled and distorted so that she could not balance on them, indicating that she had been confined to a small area for a long period and the hooves did not have a chance to wear down. It was heartbreaking.

I expressed my displeasure to Mr. Rizzo. I was angry that a person could be so neglectful of an animal, and over so long a period of time. He and his partner had to have been blind not to have seen the damage and obvious pain this goat was in. At that moment, I was so disgusted with and angry at Rizzo that I turned my back to him and gritted my teeth to keep from saying or doing things I'd regret.

Knowing this would not be a quick procedure, we made an appointment to try to trim the hooves under anesthesia and possibly get them back to an approximation of their normal state.

When she was anesthetized, we saw that the joints were very arthritic and could only be straightened temporarily with a great deal of tension. We now needed to cut or grind or saw the hooves into as normal a conformation as possible. It took all three of these methods to get the job done. The hoof material wasn't hard enough, it wasn't soft enough, and it clogged our instruments and tools. Finally, after two hours of methodical work, it was accomplished. As a final measure, we gave her a long-acting injection of corticosteroid to help the inflamed joints.

When Rizzo came for her some hours later, she was able to approximate the act of walking for the first time in many months. I gave him instructions on care and exercise, and asked to see her periodically for injections and hoof examinations before they got so out of control again.

I have a feeling that most vendors of food animals view them as inanimate objects rather than living beings, and treat them carelessly. I was angry and disappointed, but resigned. This was not apt to change. I never saw the goat again.

* * * * *

The third goat visit was the saddest. She was brought in by Jim, a former partner of Tony Rizzo who now had a chicken market in another section of Brooklyn. He had a couple of pet goats too—or, in retrospect, perhaps he was slaughtering the occasional goat as well as the chickens.

"Doc, Daisy's been pregnant for a long time. She was due to have kids about two weeks ago. Now she's acting sick. Lay down, stopped eating and is breathing funny. Can you help her?"

The goat looked down-and-out, possibly near death. Her eyes were dull, she had a subnormal temperature, pale gums and mouth, and her abdomen was incredibly distended. She resembled Picasso's sculpture of the pregnant goat.

"Why did you wait so long to bring her in? Her chances of surviving are extremely poor. She's not at all dilated. She is almost dead now." I was very upset and angry and let him know it, for all the good it would do.

"Do whatever you have to, Doc. We kept thinking day after day she'd have them and she looked and acted fine up to a couple of days ago. I'm sorry."

Sorry never makes up for neglect or changes anything.

We started Daisy on supportive treatment to get her to a point where anything could be undertaken; she was put on an intravenous drip of fluids, given antibiotics, and laid on blankets with several electric heating pads under them to raise her body temperature. The following day, our surgical assistant prepared her for a Cesarean section, which my associate Dr. Rich Turoff, and I proceeded to do.

There were four full-term kids in the two horns of the uterus. The usual number is two. All were dead and surrounded with foul, putrefying placental tissue and pus. The entire uterus with all its contents was removed, surgery was completed, and supportive treatment instituted once again. Unfortunately, Daisy was so toxic that she died after lingering two more days.

Goats are very gentle creatures and we had grown very fond of Daisy. It was a depressing day for all of us involved in her treatment, and epithets flew about periodically throughout it, aimed at the so-called caretakers of this animal. This was just another example of the lack of caring of some of those who deal in animals as chattel.

* * * * *

A client whose dog I had been treating asked if I'd be willing to examine and treat a wolf that seemed to have an ear ailment. A friend of hers was traveling the country and appearing on television interview shows with two male wolves, Tim and Jethro. The purpose of the tour was to illustrate the particular ways wolves show affection, which are generally misconstrued as dangerous to those who don't know. Wolves are intelligent animals.

From the way she asked me, I believe she thought I might be reluctant to treat him. But I was thrilled at the prospect of it! We made an appointment for several days later to see him at the end of office hours so he would not encounter any cats or dogs. I arranged to signal my family in our apartment above the office when Jethro arrived. I knew they'd want to see him too.

Jethro's owner, a muscular young man who was an ardent wildlife preservationist and environmentalist, brought him in. The animal resembled a large silvery German shepherd. Intelligence shone through his light brown eyes, and he didn't resist any more than most dogs when I examined his ear canal with an otoscope. I was not afraid to treat him since I felt that the people who restrained him were competent, and I sensed that he knew he was being helped.

I flushed and cleaned his infected, malodorous ear canal thoroughly and instilled antibiotics. I managed to give him a general examination and he appeared fine. Even got a look at his teeth, which were clean and exactly like those of my canine patients. I dispensed medication to his owner (do we ever really own an animal, especially a wild one?) and they left. It was a thrill for all of us to have been that close to a feral creature.

A month later, an article appeared in one of the New York City newspapers about a demented woman in Coney Island who had thrown poisoned meat into a specially fitted van that housed two wolves. Both were killed. They were Tim and Jethro.

* * * * *

Sal Russo kept a small pony with his large German Shepherds in a yard adjacent to his home in Gerritsen Beach. One of the dogs decided he'd like to taste some pony and bit a piece out of the base of the pony's tail. So Sal, a good-natured, humorous young fellow, always sporting a smile, put the pony in the back seat of his car, and seated like a human passenger, drove him to my office along the Brooklyn-Queens Expressway. He laughingly described the astonishment and double takes by other motorists passing him on the road when they casually glanced into his car.

Fortunately, traffic was not too heavy at that hour, which accounted for a lack of collisions and pile-ups.

We cleaned and sewed the wound and sent Sal on his way, he promising to keep the species separate.

* * * * *

Meg Merrill is an animal lover, pure and simple. A large part of her time and energy has always been devoted to helping stray pets find homes and to various humane causes. Meg had two cats that were regular patients, a long-haired gray and a semi-long-haired grey and white.

Then she obtained a margay kitten. Margays are wild, leopard-like animals indigenous to South American jungles. At the time she got the kitten, there was an extensive trade in exotic animals, especially in large port cities. They were trapped illegally in the jungles, the poachers often killing the mother to steal the litter of kittens, and then smuggled into the United States. In the long sea, air and ground trips, including layovers, many died en route.

Dealers view them as salable commodities and nothing more. They could just as well be pieces of lumber. At the time, there were no laws against owning exotic wild animals.

I don't think Meg would have taken one on if she hadn't seen it in a pet shop, shaking, skinny and malnourished, altogether sickly and forlorn—a candidate for an early death. But take him she did. He was five weeks old, weighed one and one quarter pounds and was a mess.

It took lots of love, care and good nutrition to bring this male margay, whom she named Montezuma, Monte for short, back to good health and growth to the beautiful 15-pound sinuous jungle beast he was to become.

Meg brought Monte regularly for examinations, vaccinations, and laboratory work, to check for parasites and urinary infections. The day-to-day raising of the margay was both a trying and rewarding experience, by way of observing his habits, and trying to approximate the diet that he would have had in his native habitat.

Meg had ceiling-high bookshelves and Monte loved to perch on the highest shelf, as he would have on tree branches in the jungle. He was very playful and snuggled up to sleep with Meg's domestic cats, fitting into the family.

It was no simple, short-time task to get this wild beast to integrate with a human family and live indoors. I'm sure Meg could tell tales of broken china and crystal and of multiple scratches and bite wounds inflicted on her, her husband and their son. But Monte grew up partially civilized. He was even trained to use the toilet bowl! Even though he thrived, it took an all-consuming dedication of time and energy and was still not wholly fair to the animal. To this day, Meg vehemently advises against it.

At 13 years of age, he became less lively, his appetite was off, and his water intake increased markedly. Laboratory tests indicated that his kidneys were deteriorating, a common ailment in aging cats.

When Monte was 16, Meg invited my wife and me to dinner at her Greenwich Village apartment to observe him in his unnatural habitat. He was still somewhat active, although it was evident that he did not have much longer to live. I had only seen him in the clinic to treat him, up to this time. We tentatively petted him as he sniffed us. Wild animals are gentle with their owners (up to a point) but outsiders need to be wary.

Once he was further acquainted, he bounded about the room, ending near the high ceiling on the top of a bookshelf, one paw hanging down. The evening was especially memorable for the chance to see and touch this wild animal at home.

In addition, friends of Meg's, Irv Renard and Jacque Dean, regaled us with stories of the two bobcats they had raised as cubs. One evening, Jacque told us, she had set their table for a very elegant dinner for eight guests. The bobcats, Bailey and Lumps, were playfully chasing each other around the house. The guests had just sat down at the table and wine was being poured when suddenly the animals decided to use the table as an expressway. Bailey leaped on the table, running across the length of it, followed by Lumps. One can imagine what havoc two rambunctious wild cats weighing about 30 pounds each, with huge paws, could wreak in this situation. Fine crystal wineglasses and water goblets went flying in all directions, much of it shattering on the floor, followed by silverware and china place settings shortly thereafter. Irv, Jacque and their guests were splashed with wine and appetizers. Needless to say, dinner plans were delayed pending a cleanup of the colossal mess.

Where were Bailey and Lumps after this disaster? Why, they kept up their game of tag, oblivious to all the furor they had caused.

One wintry night, when a heavy snow was falling and several inches had accumulated on the ground, Irv and Jacques' front door was accidentally left slightly ajar. A gust of wind blew it open and out bounded two innocent, playful bobcats onto Ninth Avenue in New York City. The cats were intrigued by the falling snowflakes, trying to eat them as they fell, all the while racing down a street replete with speeding cars, pedestrians, traffic lights and policemen. Upon discovering the open door, the panicked couple raced out in their underclothes after the bobcats, trying to reach them before Bailey and Lumps could be destroyed by automobiles or shot by the police. Luckily, the bobcats stopped to roll around in the snow. Irv and Jacque caught up with them and led them home by their collars.

I asked Meg later on what had finally happened to these animals. She told me that at 15 years, Lumps died of renal failure. Bailey died a year later, developing fluid in his chest cavity. Jacque was devastated by this loss for a long time.

Another feature marked the evening at Meg and Si Merrill's apartment. Bernice, who had very close contact with many cats over a long period of time, was never allergic to any of them. That night, after handling the margay, her eyes itched intensely as the evening wore on, so much that she could hardly suppress the urge to rub them. Then the tissues surrounding her eyes swelled until they were two slits and she could barely see! We left earlier than we had intended, and I led Bernice to a taxi. Antihistamines and a couple of days away from Monte and she was back to normal. Margays obviously have different protein in their dander than regular domestic shorthairs. One never knows

A few months after our dinner, Monte became very uremic. His kidneys failed. Meg had him put to sleep—the oldest documented margay raised in captivity.

* * * * *

Another margay was brought in by Mrs. Mulligan. She had bought him from a pet shop, but unlike Monte, he was in good condition. He was about four months old and a bit wild. We examined him, vaccinated him and checked for parasites. All was good.

Several months after her first visit, she arrived at my office without an animal, and the left side of her head was bandaged. Her husband was working nights at the 1964 New York World's Fair, piloting a hydrofoil and taking people for rides on Jamaica Bay. She was unable to sleep while her husband was out, so she resorted to taking sedatives. One day her husband came home near dawn to find her lying on the floor in a pool of blood. Her left ear was chewed off! She had taken a larger dose of sleeping medication than usual and had slept through this auricular feast. We all knew who did it, but Mrs. Mulligan insisted it could have been her other cat, a sweet, gentle domestic short-haired pussycat. By this time, there was a law against keeping exotic animals within the New York City boundaries, and the ASPCA ordered her to get rid of the margay. She asked me to write a note to the ASPCA stating that her other cat was the culprit! When I refused, she threatened to take her margay up to Westchester, a county north of New York, and would sneak him back at some future date. Talk about masochism! All for a margay named Jaws.

* * * * *

My practice accepted injured wild birds that people brought in from the street. We tried to repair them without charge and release those fit to fly. For the birds that remained lame or unable to take to the air, we were lucky to have a few sources that took them in as pets. Though our technician Bill was extremely allergic to pigeons, he insisted we keep up the policy of caring for them.

On one occasion, Dr. P.J. Field inserted a stainless-steel pin into the badly broken wing bone of a pigeon. We removed the pin four weeks later, and released the bird into the street. Our whole staff cheered as it flew off to a tree a good distance away.

Treating these helpless, ownerless creatures is a very satisfying thing to us all, so it was very gratifying, one day, to receive the following letter. Made my day.

The Diocesan Church of Saint Ann
and
The Holy Trinity
122 Pierrepont Street
Brooklyn Heights, N.Y. 11201
—
TRIANGLE 5-6960

Dear Dr. Wasserman:

Just a note of thanks for taking in one of the injured creatures of the earth last week. It was quite a scene on Pierrepont Street with a Priest, a plumber, an electrician and two women of undetermined vocation standing around a pigeon suffering from a broken wing trying to decide what to do with it, when suddenly one of the women called out "Wasserman"—rather like someone calllling out "mud" when a country boy is stung like a bee or calling out "water" when the occupied life raft washes up on shore. Once again, thanks—it's actions like these which, as we know, make the quality of life in New York City decent.

Yours,

Tom Faulkner

Father Thomas Faulkner

Above: Bill Hinkle with Caprine patient.
Below: Montezuma and his reflection.

Above and below: Montezuma with a domestic friend.

Taste Test

A client called one day and proceeded to ask me numerous questions about dog food: what did I think about certain brands? Did I recommend dry food over one of the most popular brands of canned dog food? Which is more nutritious? Which tastes better? I presumed he questioned me in order to offer his pet, a female Saluki named Cleo, a nourishing menu.

But the owner was Raymond Sokolov, the food editor of *The New York Times*, and he had thoughts in mind beyond a good diet for his dog. Not long after the interrogation, he featured an article discussing the merits of canned dog food vs. semi-moist and dry. Among his resources were Dr. Albert Jonah, a veterinarian heading the Animal Care Division of the Yale Medical School, and a Mrs. O'Keefe, a spokesperson for the industry-sponsored Pet Food Institute. Dr. Jonah and I concurred that an all-dry-food diet was nutritionally best, either alone or with a small amount of wet food to enhance the taste. Mrs. O'Keefe disagreed: she didn't think the enthusiasm for dry food was well founded and believed canned and semi-moist foods were more palatable.

But the main thrust of his article was a taste test in which both he and Cleo participated. To prepare, both fasted for 16 hours prior to the tasting. Cleo passed on the two dry foods, but ravenously ate all nine other varieties of foods—moist (canned), semi-moist and dog biscuits. Then the food editor sampled all the dog foods and rated them from one to four stars, depending on how they compared by taste to human foods. His comments, with a picture of Cleo, appeared in the article as follows:

To Human Critic, No Brand Had a 4-Star Taste
★★★ *Ground Chuck.* Needs seasoning.
★★★ *Milk-Bone Biscuit.* Could replace Ry-Krisp with a little salt and butter.
★★ *Prime, chicken flavored.* No chicken taste: moist, sweet cubes like yellow cake.
★★ *Medallion, beef-flavored chunks.* Texture like cake, a strong meat flavor.
★ *Purina Dog Chow.* Stale biscuit texture, but subtle meat flavor; not appreciably dry when moistened.
★ *Recipe, beef and egg dinner.* Excellent odor, like chop suey; mushy texture and no seasoning.
★ *Laddie Boy, lamb chunks.* Best odor of all moist foods, but no taste, gooey texture.
Top Choice, chopped burger. Tasteless, rubbery, drastically red color, pasty in mouth.
Gaines Meal. Like concretized sawdust.
Alpo Horsemeat Chunks. Awful-looking, smelled like stew, tasted foul.
Unrated: *Daily All-Breed, liver flavor.* Strong, mysterious odor, couldn't get it down.

To Human Critic, No Brand Had a 4-Star Taste

*** **Ground chuck.** Needs seasoning.

*** **Milk-Bone Biscuit.** Could replace Ry-Krisp with a little salt and butter.

** **Prime, chicken - flavored.** No chicken taste; moist, sweet cubes like yellow cake.

** Medallion, beef-flavored chunks. Texture like cake, a strong meat flavor.

* **Purina Dog Chow.** Stale biscuit texture, but subtle meat flavor; not appreciably dry when moistened.

* **Recipe, beef and egg dinner.** Excellent odor, like chop suey; mushy texture and no seasoning .

* **Laddie Boy, lamb chunks.** Best odor of all moist foods, but no taste, gooey texture.

Top Choice, chopped burger. Tasteless, rubbery, drastically red color, pasty in mouth.

Gaines Meal. Like concretized sawdust.

Alpo Horsemeat Chunks. Awful-looking, smelled like stew, tasted foul.

Unrated

Daily All-Breed, liver flavor. Strong, mysterious odor, couldn't get it down.

Bennington

What a cat! Bennington is one of the most remarkable and entertaining felines that ever existed. Sure, you say, all cat owners think that way about their pets. But Bennington really is among the greats.

A slim, delicate seal-point Siamese named Alexander had been wandering the streets, frightened and bewildered. We tried to locate the owner, without success. From his coloration, head and body structure, I was sure this cat was of a strain of Siamese that was bred by a client named Bill Arnholt. Bill verified this but couldn't help determine the owner.

My wife and I decided to keep him. For the seven years we had Alexander, I was never able to pick him up. He allowed me to pet him, but only tentatively. My wife fared somewhat better. Nevertheless, he was gentle and humorous, and at times, dragged his petite Siamese companion, Homerella, all over the house by the scruff of her neck. When he died suddenly, we were devastated. On several occasions when a seal-point Siamese came in for a visit, I had to suppress a tear remembering Alexander.

And so it was, a couple of months later, when the staff of the veterinary clinic presented me with a five-week-old Siamese kitten for my birthday. Bernice and I named him Bennington after the county in Vermont where we vacationed, and to keep the name close to Benjamin, our wonderful cat of the past.

Bennington at first was a tiny male seal-point who fit into the palm of my hand, not much bigger than a mouse. However, he needed no coaxing to eat and eat and eat and it seemed as though one could watch him grow as he ate. From the start, he exhibited the characteristics of the cat he was to become: independence, intelligence, friskiness, playfulness, affection, and fierceness. As almost all cats do, he trained himself immediately, seeking out the litter box. He slept on the bed with us, and, as small as he was, made his way off the bed to the bathroom and litter box, and climbed up the spread on his return using his claws.

One morning, I reached for him while he was half-awake. He lashed out and scratched my eyeball, barely missing the transparent part of my eye called the cornea. Though the scratch was painful, medication prevented infection, and I learned not to surprise him. I also learned there were parts of his body he didn't like handled, mainly his abdomen—he'd go for me if I dared unless he was on my lap, belly up. Then he loved it, and the harder I rubbed or paddled, the better! Go figure.

With Bennington, one has the sense of a wild animal living in the house, a creature from the forest that elects to bestow friendliness and affection. Otherwise, he loves to be roughed up, to have his back and sides violently massaged, all the while purring continuously, and then he begs for more. I've even gotten him to flip from one side to the other several times in sequence for more rubbing—but of course he'll seldom do it when I want to show him off to others.

When Bennington was a few months old he bounded onto a chest of drawers and knocked off all sorts of objects, including a number of Wedgwood china pieces. He batted the crystals hanging from our two Waterford sconces, eventually knocking them off. We called him the thousand-dollar cat. We've raised his price since, though he hasn't broken anything in several years.

And by the way, have I mentioned how beautiful he is? He is large, especially as Siamese cats go (he grew to be 18 pounds). He is beautifully proportioned and colored with dark seal-points on his face, ears, paws and tail. His body is tan and his eyes the color of the Caribbean—not the sapphire hue of the water closest to shore, but the darker blue water near the horizon which is tinged with violet.

Bennington loves contact with people. He ambles up to us, or even strangers, with a John Wayne swagger, squeezes in beside the person on a chair and lies contentedly while they're talking or reading. A few strokes on his back or head will bring forth a continuous drumroll of purring. Or he'll lie down alongside someone with just his paw in contact—or he'll drape himself into the crook of one's arm with his head and forelimbs on their chest. He loves to be in touch.

With all other cats, he shares the trait of sitting on top of all sorts of papers I am trying to work on—bills, letters, taxes, etc.—while purring contentedly and gazing at my face. It is both maddening and endearing. How can I be angry with my cat for wanting to be with me? I can't. Additionally, like most cats, he will sleep all day and then declare "playtime" at night by racing around the bedroom, batting pictures on the wall with his paws, and jumping on and off the bed until I must lock him out of the room, only to have to endure howling outside the door. Not someone to be ignored, my cat!

He has an endearing habit of luring me away from whatever I'm doing by engaging in his picture-bouncing calisthenics, and when I come after him, throwing himself at my feet so I can rub him up fiercely—the harder, the more ecstasy he's in.

My Bennington will jump on the table whenever we sit down, meowing in our faces, tapping our arms with his paw, touching and pushing his packet of Tender Vittles until I hand feed him this treat. He doesn't know when to stop. My needs and my desires come last. Instant gratification is his game.

His amazing range of sounds varies from a cute, eager, happy meow to his sweet squeak upon being picked up, to a short sound that can only be described as an exclamation point—different from his question mark. A few times we've actually heard a roar from him that should have emanated from an animal ten times his size. The first time we heard it, we were astonished! We thought a lion had entered the house. In her recently published book *The Tribe of Tiger*, Elizabeth Marshall Thomas says that cats are divided into the categories of big cats and small cats. This division is not altogether based on size. For instance, the clouded leopard, a small feline, is in the big-cat column, while the puma, a large animal is counted with the small cats! What is the main criterion for differentiation? Large cats are able to roar. Small cats cannot! Where does that put my pussycat? He is an enigma.

Bennington's facial expressions could put some actors to shame: sleepy-faced (hey, who turned the light on?), imploring (come on, scratch my back), affectionate, angry (watch out if his ears are folded back as well), mild annoyance (aw, stop smooching me already), and many other subtle faces in between that we have learned to read.

A superb athlete, you can almost sense the computer in his head measuring distances and heights as evidenced by tiny movements of his toes, and his steadfast gaze at the target as he prepares to jump. Bennington never misses landing perfectly and gracefully on the smallest surface. At the prompting "go for the gold," he will enthusiastically leap six feet from a chest of drawers onto the bed. Otherwise, there he would remain indefinitely causing havoc with objects on the chest. We started prompting him during the last Olympic Games and he responded to it. Thinks he's in the broad-jump competition. At the completion of the jump, he gets cheers, accolades, and hugs from us, which he seems to look for. He's a ham.

When we spend time in our country house, Bennington assumes an entirely different personality. Aside from the occasional morning stroll through the tall grass, he spends most of his time in an upstairs bedroom under the bedspread, so that he is a bump meowing upon hearing a noise in the room. He seldom comes downstairs and only ventures from his nest to eat and use the litter box, but he is not unhappy.

None of my cats seems interested in mice or birds. I have seen mice boldly eat from their cat food dish and evoke not much more than a yawn. It is beneath Bennington's dignity to harm such a wee, helpless creature. Jenny shows a bit of animation, mostly an expression of curiosity, but she doesn't harm mice either.

On the other hand, Bennington will take on larger species like Bill Hinkle, our veterinary technician. Bennington wants to kill Bill.

Once, long ago, Bill agreed to feed and water our cats daily in our absence. When Bill came through the door, Bennington leaped at his throat. Using a chair as a fulcrum, he tried to get at Bill's jugulars and carotids. Bill was very shaken. He ran out of the door and came back with a stout stick to ward off Bennington while he filled the cat's dishes until Bill found a substitute feeder. Such attacks did not occur with any other cat sitters. Weeks later, hoping Bennington had gotten over his grudge, Bill came in while we were home—but our jungle cat once again attacked savagely, this time going for Bill's ankles.

What was it about Bill that singled him out for attack? Until this cat is ready to talk—and I think he has the intelligence to learn—we'll never know.

Bennington avidly grooms my hair with his raspy tongue at least once a day, usually at night as I lay down. I often see Jenny and him groom each other. I think he has accepted me as part of his tribe.

Sometimes Bennington and I see into each other's eyes and I communicate how much I appreciate and enjoy him. I can only hope I mean as much to him.

Bennington turned 15 not long ago. The thought of losing him in a few years is very sorrowful for us, so we make a point to enjoy his presence day by day.

A Frenchman, Méry, wrote, "God made the cat to give man the pleasure of caressing the tiger." Bennington is our tiger. Or lion.

* * * * *

Two and a half years have passed since the portrait above was written. Bennington has since died.

At about age 16, he started to develop symptoms of kidney deterioration: loss of weight, copious thirst and diminished appetite. Periodic blood chemistries indicated that his kidneys were getting worse with time. Only a change of diet could have slowed down the process, but he would not eat the beneficial foods. Rather than let him starve, he stayed on a less than optimal diet.

He continued all his mischievous, humorous and lovable ways, still up on the table with his face in yours, looking for a handout of his favorite food. He even did it when he didn't want any. It had become a routine, a thing to do, a game. He became increasingly emaciated.

Bennington still sought me out if I sat reading on a reclining chair or sofa and wedged his frail body between me and the chair arm. It was wonderful to have him there, to have the contact with him, lying contentedly. He still slept in the crook of my arm or draped around Bernice's neck.

Though less frequently, he still pulled his stunt of getting up on the piano, putting a paw behind a picture and thwacking it so it bounced off the wall and then, when confronted by my mock anger, jumping off the piano and throwing himself on the floor to be rubbed vigorously.

I don't know how many times we hugged and kissed him every day, knowing his days were numbered, and how thankful we were that, unlike humans, animals don't know they are going to die.

I didn't mention in the portrait what a naturally perfumed odor and taste his body had, and rubbing my cheek over his fur was a treat. He was just plain delicious.

Bennington became deaf at age 16½. We realized that he didn't respond to our voices when he wasn't facing us. This necessitated our seeking him out when we wanted to feed him—or smooch him.

In his final days, when he was too weak to walk, I gave him subcutaneous fluids daily and oral nutrients in gel form. He slept on our bed on a soft sponge mattress. During the first night, he cried out and we comforted him; he really was trying to say that he needed to urinate, which was obvious in a moment. After that, his nocturnal cry was a signal for me to carry him to the bathroom and express his bladder over the bowl. He was totally limp. He lingered on for two weeks, and one night, when my son Andy, his wife Donna and our grandson Jake were visiting, we sensed that Bennington was going to leave us. We all kissed him, including 20-month-old Jake who had romped with Bennington six months earlier. Bennington closed his eyes for the last time on September 10, 1997.

Two days and rivers of tears later, we took him to our country home and buried him under a majestic maple tree.

An amateur potter, I have made a ceramic plaque to be put on a granite or marble marker. It reads:

BENNINGTON
January 1980 - September 1997
Forever Missed
Always Loved
A Cat for All Seasons

He was seventeen years, eight months old.

Bennington, age 13

Above left: Jenny.
Above right: Jenny as a kitten.
Below: Bennington, age 8.

Living Dangerously

Part of the Brooklyn docks is in an area called Red Hook, where the Atlantic Ocean, passing through The Narrows, meets the waters from the Hudson and East Rivers.

Large ships from all over the world dock in Red Hook and unload, their international cargoes transferred into trucks and railroad cars for further shipment.

Red Hook adjoins my neighborhood and is a mostly blue-collar area of nice, law-abiding citizens. Some of my clients from Red Hook worked as longshoremen, some worked for the union of dock workers, and a few did who-knows-what in that shady underworld group known as the "mob." The area is not unlike the docks seen in the movie *On The Waterfront.*

One day, a man brought a part-German Shepherd puppy to the office, with a classic case of distemper: caked-up nasal discharge of mucus and pus, a similar discharge affecting the eyes, a severe cough, diarrhea, loss of appetite and weight, and poor coat. He was mighty sick.

When making out the pup's record card, we had trouble eliciting the name of its owner. He was the issue of the "club" mascot, the "club" being a storefront hangout of some of the less public-minded individuals of the area. The man presenting the dog insisted on giving the club as owner and I insisted on one person's name. He finally gave me his, "John Sacco."

We hospitalized the pup, treated him with fluids and antibiotics and within three days, he seemed fine.

Distemper is a strange viral disease. Generally, about half of the stricken dogs either die or end up with severe neurological symptoms, which may necessitate euthanasia. Sometimes they appear to make a full recovery, only to have a relapse and then either recover or go completely bad.

When we sent the pup home, a person other than Sacco called for him and paid the modest bill. The puppy was full of spirit when he left. Alas, we learned several days later the spirit didn't last long.

Once again, he was admitted. We told the person who delivered him not to expect too much.

This time the puppy made slow but steady progress, and two weeks later, we were reasonably sure that he was one of the lucky ones.

We called the club constantly, each time getting a different person, and asked the person to pick up the pup and pay the fee. Each time, the person said he'd come, and each time failed to do so.

I knew it couldn't be the fee since it would have been divided by the club members—and besides, it was a modest fee. Looking back through some record cards from that time, many years ago, I see that an office visit was $3.00! A visit with a penicillin injection and some pills came to about $5.00. A day's hospitalization amounted to not much more.

Two more weeks passed, and Kelly, the pup, was thriving: rambunctious, growing rapidly, eating us out of existence as only a young, large-breed puppy can do. I had to get him out. It was unfair to keep him continuously kenneled. Though he was a lovable dog, he barked incessantly. He wanted to be out running and playing.

It appeared that Kelly was going to be a "hangup"—a

patient whose owner abandons him for whatever reason. At that time, there was no mechanism for dealing with abandoned animals. Since then, a state law has been in effect enabling the veterinarian, after sending a certified letter to the owner and receiving no response, to turn the pet over to the American Society for the Prevention of Cruelty to Animals. Since this predated such redress, I turned Kelly over to my friend Bernie Gordon who owned a pet shop around the corner. Bernie promptly sold him for $25.00, my $12.50 share being a fraction of what we had spent.

End of story? Not by a long shot. Two weeks later, the telephone—a gruff man's voice.

"Eh—I'm calling about da Sacco pup. Ow's he doin'?"

I was shocked! The pup was irretrievable.

"He's fine. In great shape," I said nonchalantly. Then I decided to bluff and push this to a conclusion. "Why do you keep calling but never collecting the pup? Either come in or forget about it." I hoped they would do the latter.

"How much is da bill?" More than two months had passed since the pup had been dropped off, and I quoted the caller a pretty hefty fee (for those times). I got a "wow" and he hung up.

Now what? Were they really still interested? What would I do if they took up a collection and decided to retrieve the pup? Though logically, I had the right to take the action I did, who knows what could have ensued? Frankly, I was scared, and a cloud of anxiety hovered over me day and night.

Nothing for several more weeks—and then a woman's voice on the telephone.

"Hello, I'm calling about Kelly. How is he?"

I almost dropped the receiver.

"He's great," I said, trying to sound calm and cheerful. "What does the bill come to?"

I quoted her a fee, now considerably higher than the last one.

A gasp, from the other end of the line. "We can get a purebred German Shepherd puppy for much less than that," said the voice. Apparently no great attachment to this particular pup.

"You certainly can," I replied.

I sensed this long drawn-out case had reached its conclusion. And it had—except that a few months later, a picture on the front page of *LIFE* magazine showed the capture of a notorious Brooklyn mobster (who subsequently served a jail sentence, and then, after his release, was gunned down eating clams in a Manhattan restaurant) along with his bodyguard. It was Sacco, the man who first brought the pup in.

The Dog Who Met the Queen

It is relatively infrequent that a human being appreciably changes the circumstances of his or her life. We are born into a certain kind of environment and parentage, and what is attained or accomplished may be circumscribed. There are exceptions, of course. Through education, or winning the lottery, or entrepreneurship, some break out of the mold and arrive at a different plateau of life.

Animals in the wild have no chance of changing their destiny. They live out their lives in forests, mountains, deserts and plains just struggling to exist—to kill and devour for food, or to keep from being killed and devoured as some other animal's meal. Domestic farm animals raised for food are, of course, doomed from the start.

But consider the world of the dog and cat. Depending on pure chance, a puppy or kitten could end up euthanized due to overpopulation of pet animals. They could wind up as street strays with all the hazards this entails: hunger, suffering from cold or heat, dodging (not always successfully) automobiles, and cruel treatment at the hands of people. They could be bought or adopted by people who may mistreat and later abandon them.

On the other hand, many a dog and cat has landed in a good home with caring and doting owners, to live out a happy, carefree life. Veterinarians mostly see this category of pet; the others are given a minimal amount of care, if any. The fortunate pets are mourned when they die and are given burials or cremation. Such animals are literally members of the family.

One outstanding example of a dog that lucked out belonged to a client, Caroline Cutler. She went to the A.S.P.C.A. shelter and adopted a nine-month-old female of very mixed parentage, to live with her and her young son Thomas. The dog was of medium size, grey and white, and shaggy—the type often referred to as a Walt Disney dog.

Caroline was a warm, vivacious woman, dark-haired, dark-eyed and very petite. Her friendliness and deep South Carolina drawl were contagious. I knew that Topsy, as her new pet was called, was one of the fortunate ones.

Caroline had indeed picked the neediest dog there. Topsy was emaciated, with hair so matted and greasy and overgrown that her eyes were not visible. She cowered in fear and mistrust. She was a sure candidate for euthanasia, a solution for unadopted dogs at many shelters. But left in Caroline's loving care, Topsy thrived. She became a happy, playful and affectionate dog.

On a visit several years later, Caroline brought Topsy in for a rabies vaccine injection and a health certificate preparatory to shipping the dog abroad. She told us she was to marry a man in the diplomatic service of the U.S. Government, on post to London. Topsy had to be quarantined for six months due to the same strict international laws. Caroline arranged to have her kept near London so she could visit her dog regularly until the quarantine was over. Then Topsy would come and live with her owners in a magnificent home allotted to the diplomat second in command of the United States Embassy.

She issued warm goodbyes to me, my associate and the

rest of my staff. We had established a friendly relationship—not hard to do with Caroline.

I mentioned that my wife and I go to London fairly regularly; Caroline gave me her address and said to be sure to contact her on our next visit.

On one trip to London, my wife and I were invited to tea and saw Topsy and Scarlet, a female yellow Labrador Retriever puppy. Scarlet was keeping Topsy active and fit with her constant puppy exuberance.

In due course, Caroline's husband was reassigned back to Washington to serve at the State Department and later returned to England as Ambassador. Now, two dogs had to be quarantined for six months! While waiting for their release in a petless home, they adopted a third dog, an older Golden Retriever named Chloe. All three dogs now lived in a stately mansion.

Subsequently, Bernice and I were again invited to tea by Caroline. Her husband was lecturing in the United States at the time. Winfield House is the United States Ambassador's residence and is located in Regents Park, a large and lovely area of greenery in London. Caroline gave us a tour of the stately house and gardens and we shared an enjoyable conversation about Brooklyn Heights, London, and the interesting society of writers, poets, actors and statesmen who had been to receptions at Winfield House.

Topsy walked very slowly, showing signs of age, but as sweetly as ever, escorting us from the guard's gate to the house when we arrived. And we saw and experienced the exuberant, rambunctious and playful Scarlet who loved to chase tennis balls. Scarlet's balls were scattered all over a large banquet hall, filled with priceless antique chests, sideboards and other treasures. Scarlet ran into them all while retrieving the balls we threw, often returning with three in her mouth at the same time, looking very cute and silly.

Stately mansion or not, this was her playground. When we left, Chloe walked us back to the guard's gate.

Not long ago, we received a letter from Caroline. In part, it said:

"I don't know if you remember what Topsy looks like (how could you forget?), but it should amuse you greatly to know that she, scruffy little abandoned dog, rescued from the pound in Brooklyn, met the Queen when she came here to dinner! Talk about a Cinderella story."

Can you imagine Topsy at an animal shelter-adoptee reunion? What a tale she could tell!

About the Author

Bernard Wasserman, AB, D.V.M. graduated cum laude from the veterinary school at Michigan State University in 1951. For the succeeding three years, he was an Assistant Professor of Animal Pathology at the University of Rhode Island. There at the Agricultural Experiment Station, he and an associate ran a diagnostic laboratory for animal diseases, conducted research on viral diseases and taught in the College of Agriculture.

He turned to the practice of small animal medicine in 1954 and opened his own small animal hospital in 1957. For the fun of it, he started writing a column on pets for a neighborhood weekly newspaper, *The Brooklyn Heights Press* in 1958. The columns caught the eye of *The Herald Tribune Syndicate*—and the editor went to the doctor—and the "Vet on Pets" column was born.

Retired from practice, he pursues his hobby as a studio potter, travels frequently, and indulges in reading, music and museuming.

He is married to Bernice. They have two sons and one grandson, Jake.